REVIEWS OF
MODERN PHYSICS

BOARD OF EDITORS

JOHN T. TATE, *Editor (on leave)* J. W. BUCHTA, *Acting Editor*

Associate Editors

A. H. COMPTON K. K. DARROW E. C. KEMBLE
K. T. COMPTON D. L. WEBSTER

VOLUME 15
1943

Published for the

AMERICAN PHYSICAL SOCIETY

by the

AMERICAN INSTITUTE OF PHYSICS
Incorporated

REVIEWS OF MODERN PHYSICS

Reviews of Modern Physics is published quarterly to provide those interested in developments in physics with comprehensive and timely discussions of the problems which are of especial interest and importance.

It is hoped that the studies in this journal will give to the specialist the benefit of the viewpoint of other specialists; to the student, a background of critical knowledge and an adequate bibliography of source material; to those whose time is absorbed in the teaching of physics, a stimulating account of the forward movement of their science.

Manuscripts for publication should be submitted to J. W. Buchta, University of Minnesota, Minneapolis, Minnesota.

Proof and all correspondence concerning papers in the process of publication should be addressed to the Publications Manager, American Institute of Physics, 175 Fifth Avenue, New York, New York.

Subscription Price

	U. S. and Canada	Elsewhere
To members of Founder Societies of the Institute, of the American Chemical Society and the American Mathematical Society	$3.00	$3.40
To all others	4.00	4.40

Back Numbers

Complete set: Vol. 1, July, 1929–Vol. 14, 1942—**$54.00**
Yearly rate: **$4.40**
Single copies: **$1.20** each.

Subscriptions and orders for back numbers should be addressed to Prince and Lemon Streets, Lancaster, Pennsylvania, or to the American Institute of Physics, 175 Fifth Avenue, New York, New York.

Changes of address should be addressed to the Publications Manager.

Reviews of Modern Physics is published quarterly in January, April, July and October at Prince and Lemon Streets, Lancaster, Pennsylvania.

Entered at the Post Office at Lancaster, Pennsylvania, as second class matter.

Acceptance for mailing at a special rate of postage provided for in the Act of February 28, 1925, embodied in paragraph 4, Section 412, P. L. & R., authorized February 19, 1932.

REVIEWS OF

MODERN PHYSICS

VOLUME 15, NUMBER 1 JANUARY, 1943

Stochastic Problems in Physics and Astronomy

S. CHANDRASEKHAR

Yerkes Observatory, The University of Chicago, Williams Bay, Wisconsin

CONTENTS

APPENDIXES

INTRODUCTION

In this review we shall consider certain fundamental probability methods which are finding applications increasingly in a wide variety of problems and in fields as different as colloid chemistry and stellar dynamics. However, a common characteristic of all these problems is that interest is focused on a property which is the result of superposition of a large number of variables, the values which these variables take being governed by certain probability laws. We may cite as illustrations two examples:

(i) The first example is provided by the *problem of random flights*. In this problem, a particle undergoes a sequence of displacements $\boldsymbol{r}_1, \boldsymbol{r}_2, \cdots, \boldsymbol{r}_i, \cdots$, the magnitude and direction of each displacement being independent of all the preceding ones. But the probability that the displacement $\boldsymbol{r}_i$ lies between $\boldsymbol{r}_i$ and $\boldsymbol{r}_i+d\boldsymbol{r}_i$ is governed by a distribution function $\tau_i(\boldsymbol{r}_i)$ assigned *a priori*. We ask: What is the probability $W(\boldsymbol{R})d\boldsymbol{R}$ that after N displacements the coordinates of the particle lie in the interval $\boldsymbol{R}(=[x, y, z])$ and $\boldsymbol{R}+d\boldsymbol{R}$. It is seen that in this problem the position $\boldsymbol{R}$ of the particle is the resultant of N vectors, $\boldsymbol{r}_i$, $(i=1, \cdots, N)$ the position and direction of each vector being governed by the probability distributions $\tau_i(\boldsymbol{r}_i)$. As we shall see the solution to this problem provides us with one of the principal weapons of the theory.[1]

(ii) We shall take our second illustration from stellar dynamics. The gravitational force acting on a star (per unit mass) is given by

$$\boldsymbol{F}=G\Sigma M_i\boldsymbol{r}_i/|\boldsymbol{r}_i|^3 \qquad (1)$$

where M_i denotes the mass of a typical "field" star and $\boldsymbol{r}_i$ its position vector relative to the star under consideration and G the constant of gravitation. Further in Eq. (1) the summation is extended over all the neighboring stars. We now suppose that the distribution of stars in the neighborhood of a given one is subject to fluctuations and that stars of different masses occur in the stellar system according to some well defined empirically established law. However, the fluctuations in density are assumed to be subject to the restriction of a constant average density of n stars per unit volume. We ask: What is the probability that $\boldsymbol{F}$ lies between $\boldsymbol{F}$ and $\boldsymbol{F}+d\boldsymbol{F}$? Again, the force acting on a star is the resultant of the forces due to all the neighboring stars while the spatial distribution of these stars and their masses are subject to well-defined laws of fluctuations.

From the foregoing two examples it is clear that one of the principal problems under the circumstances envisaged is the specification of the distribution function $W(\boldsymbol{\Phi})$ of a quantity $\boldsymbol{\Phi}$ (in general a vector in hyper-space) which is the resultant of a large number of other quantities having assigned distributions over a range of values. A second fundamental problem in the theories we shall consider concerns questions relating to *probability after-effects*[2]—a notion first introduced by Smoluchowski. We may broadly describe the nature of these questions in the following terms: A certain quantity $\boldsymbol{\Phi}$ is characterized by a stationary distribution $W(\boldsymbol{\Phi})$. We first make an observation of $\boldsymbol{\Phi}$ at a certain instant of time $t=0$ (say) and again repeat our

[1] For historical remarks on this problem of random flights see the Bibliographical Notes at the end of the article.

[2] This is the translation of the German word "Wahrscheinlichkeitsnachwirkung" coined by M. von Smoluchowski.

observation at a later time t. We ask: What can we say about the possible values of $\mathbf{\Phi}$ which we may expect to observe at time t when we already know that $\mathbf{\Phi}$ had a particular value at $t=0$? It is clear that if the second observation were made after a sufficiently long interval of time, we should not, in general, expect any correlation with the fact that $\mathbf{\Phi}$ had a particular value at a very much earlier epoch. On the other hand as $t\rightarrow 0$ the values which we would expect to observe on the second occasion will be strongly dependent on what we observed on the earlier occasion.

An example considered by Smoluchowski in colloid statistics illustrates the nature of the problem presented in theories of probability after-effects: Suppose we observe by means of an ultramicroscope a small well-defined element of volume of a colloidal solution and count the number of particles in the element at definite intervals of time τ, 2τ, 3τ, etc., and record them consecutively. We shall further suppose that the interval τ between successive observations is not large. Then the number which is observed on any particular occasion will be correlated in a definite manner with what was observed on the immediately preceding occasion. This correlation will depend on a variety of physical factors including the viscosity of the medium: thus it is clear from general considerations that the more viscous the surrounding medium the greater will be the correlation in the numbers counted on successive occasions. We shall discuss this problem following Smoluchowski in some detail in Chapter III but pass on now to the consideration of another example typical of this theory.

We have already indicated that a fundamental problem in stellar dynamics is the specification of the distribution function $W(\mathbf{F})$ governing the probability of occurrence of a force $\mathbf{F}$ per unit mass acting on a star. Suppose that $\mathbf{F}$ has a definite value at a given instant of time. We can ask: How long a time should elapse on the average before the force acting on the star can be expected to have no appreciable correlation with the fact of its having had a particular value at the earlier epoch? In other words, what is the *mean life* of the state of fluctuation characterized by $\mathbf{F}$? In a general way it is clear that this mean life will depend on the state of stellar motions in the neighborhood of the star under consideration in contrast to the probability distribution $W(\mathbf{F})$ which depends only on the average number of stars per unit volume. The two examples we have cited are typical of the problems which are properly in the province of the theory dealing with probability after-effects.

A physical problem, the complete elucidation of which requires both the types of theories outlined in the preceding paragraphs, is provided by Brownian motion. We shall accordingly consider certain phases of this theory also.

CHAPTER I

THE PROBLEM OF RANDOM FLIGHTS

The problem of random flights which in its most general form we have already formulated in the introduction provides an illustrative example in reference to which we may develop several of the principal methods of the theories we wish to describe. Accordingly, in this chapter, in addition to providing the general solution of the problem, we shall also discuss it from several different points of view.

1. The Simplest One-Dimensional Problem: The Problem of Random Walk

The principal features of the solution of the problem of random flights in its most general form are disclosed and more clearly understood by considering first the following simplest version of the problem in one dimension:

A particle suffers displacements along a straight line in the form of a series of *steps* of equal length, each step being taken, either in the forward, or in backward direction with equal probability $\frac{1}{2}$. After taking N such steps the particle *could* be at any of the points[3]

$$-N,\ -N+1,\ \cdots,\ -1,\ 0,\ +1,\ \cdots,\ N-1 \text{ and } N.$$

We ask: What is the probability $W(m, N)$ that the particle arrives at the point m after suffering N displacements?

We first remark that in accordance with the conditions of the problem each individual step is equally likely to be taken either in the back-

[3] These can be regarded as the coordinates along a straight line if the unit of length be chosen to be equal to the length of a single step.

ward or in the forward direction quite independently of the direction of all the preceding ones. Hence, all possible sequences of steps each taken in a definite direction have the same probability. In other words, the probability of any given sequence of N steps is $(\frac{1}{2})^N$. The required probability $W(m, N)$ is therefore $(\frac{1}{2})^N$ times the number of distinct sequences of steps which will lead to the point m after N steps. But in order to arrive at m among the N steps, *some* $(N+m)/2$ steps should have been taken in the positive direction and the remaining $(N-m)/2$ steps in the negative direction. (Notice that m can be even or odd only according as N is even or odd.) The number of such distinct sequences is clearly

$$N!/[\tfrac{1}{2}(N+m)]![\tfrac{1}{2}(N-m)]!. \quad (2)$$

Hence

$$W(m, N)=\frac{N!}{[\frac{1}{2}(N+m)]![\frac{1}{2}(N-m)]!}\left(\frac{1}{2}\right)^N. \quad (3)$$

In terms of the binomial coefficients C_r^n's we can rewrite Eq. (3) in the form

$$W(m, N)=C^N_{(N+m)/2}\left(\frac{1}{2}\right)^N, \quad (4)$$

in other words we have a *Bernoullian distribution*. Accordingly, the expectation and the mean square deviation of $(N+m)/2$ are (see Appendix I)

$$\left.\begin{aligned}\tfrac{1}{2}\langle N+m\rangle_{\text{Av}}&=\tfrac{1}{2}N,\\ \langle[\tfrac{1}{2}(N+m)-\tfrac{1}{2}N]^2\rangle_{\text{Av}}&=\tfrac{1}{4}N.\end{aligned}\right\} \quad (5)$$

Hence,

$$\langle m\rangle_{\text{Av}}=0;\quad \langle m^2\rangle_{\text{Av}}=N. \quad (6)$$

The root mean square displacement is therefore $\sqrt{N}$.

We return to formula (3): The case of greatest interest arises when N is large and $m\ll N$. We can then simplify our formula for $W(m, N)$ by using Stirling's formula

$$\log n!=(n+\tfrac{1}{2})\log n-n+\tfrac{1}{2}\log 2\pi+O(n^{-1})\,(n\to\infty). \quad (7)$$

Accordingly when $N\to\infty$ and $m\ll N$ we have

$$\begin{aligned}\log W(m, N)\simeq{}&(N+\tfrac{1}{2})\log N\\ &-\tfrac{1}{2}(N+m+1)\log\left[\frac{N}{2}\left(1+\frac{m}{N}\right)\right]\\ &-\tfrac{1}{2}(N-m+1)\log\left[\frac{N}{2}\left(1-\frac{m}{N}\right)\right]\\ &-\tfrac{1}{2}\log 2\pi-N\log 2.\end{aligned} \quad (8)$$

But since $m\ll N$ we can use the series expansion

$$\log\left(1\pm\frac{m}{N}\right)=\pm\frac{m}{N}-\frac{m^2}{2N^2}+O(m^3/N^3). \quad (9)$$

Equation (8) now becomes

$$\begin{aligned}\log W(m, N)\simeq{}&(N+\tfrac{1}{2})\log N-\tfrac{1}{2}\log 2\pi-N\log 2\\ &-\tfrac{1}{2}(N+m+1)\left(\log N-\log 2+\frac{m}{N}-\frac{m^2}{2N^2}\right)\\ &-\tfrac{1}{2}(N-m+1)\left(\log N-\log 2-\frac{m}{N}-\frac{m^2}{2N^2}\right).\end{aligned} \quad (10)$$

Simplifying the right-hand side of this equation we obtain

$$\log W(m, N)\simeq-\tfrac{1}{2}\log N+\log 2-\tfrac{1}{2}\log 2\pi-m^2/2N. \quad (11)$$

In other words, for large N we have the asymptotic formula

$$W(m, N)=(2/\pi N)^{\frac{1}{2}}\exp(-m^2/2N). \quad (12)$$

A numerical comparison of the two formulae (3) and (12) is made in Table I for $N=10$. We see that even for $N=10$ the asymptotic formula gives sufficient accuracy.

Now, when N is large it is convenient to introduce instead of m the net displacement x from the starting point as the variable:

$$x=ml \quad (13)$$

where l is the length of a step. Further, if we consider intervals Δx along the straight line which are large compared with the length of a

TABLE I. The problem of random walk: the distribution $W(m, N)$ for $N=10$.

m	From (3)	From (12)
0	0.24609	0.252
2	0.20508	0.207
4	0.11715	0.113
6	0.04374	0.042
8	0.00977	0.010
10	0.00098	0.002

step we can ask the probability $W(x)\Delta x$ that the particle is likely to be in the interval $x, x+\Delta x$ after N displacements. We clearly have

$$W(x, N)\Delta x = W(m, N)(\Delta x/2l), \quad (14)$$

since m can take only even or odd values depending on whether N is even or odd. Combining Eqs. (12), (13), and (14) we obtain

$$W(x, N) = \frac{1}{(2\pi Nl^2)^{\frac{1}{2}}} \exp(-x^2/2Nl^2). \quad (15)$$

Suppose now that the particle suffers n displacements per unit time. Then the probability $W(x, t)\Delta x$ that the particle will find itself between x and $x+\Delta x$ after a time t is given by

$$W(x, t)\Delta x = \frac{1}{2(\pi Dt)^{\frac{1}{2}}} \exp(-x^2/4Dt)\Delta x, \quad (16)$$

where we have written

$$D = \tfrac{1}{2}nl^2. \quad (17)$$

We shall see in §4 that the solution to the general problem of random flights has precisely this form.

2. Random Walk with Reflecting and Absorbing Barriers

In this section we shall continue the discussion of the problem of random walk in one dimension but with certain restrictions on the motion of the particle introduced by the presence of reflecting or absorbing walls. We shall first consider the influence of a reflecting barrier.

(a) *A Reflecting Barrier at* $m=m_1$

Without loss of generality we can suppose that $m_1>0$. Then, the interposition of the reflecting barrier at m_1 has simply the effect that whenever the particle arrives at m_1 it has a probability unity of retracing its step to m_1-1 when it takes the next step. We now ask the probability $W(m, N; m_1)$ that the particle will arrive at $m(\leqslant m_1)$ after N steps.

For the discussion of this problem it is convenient to trace the course of the particle in an (m, N)-plane as in Fig. 1. In this diagram, the displacement of a particle by a step means that the representative point moves upward by

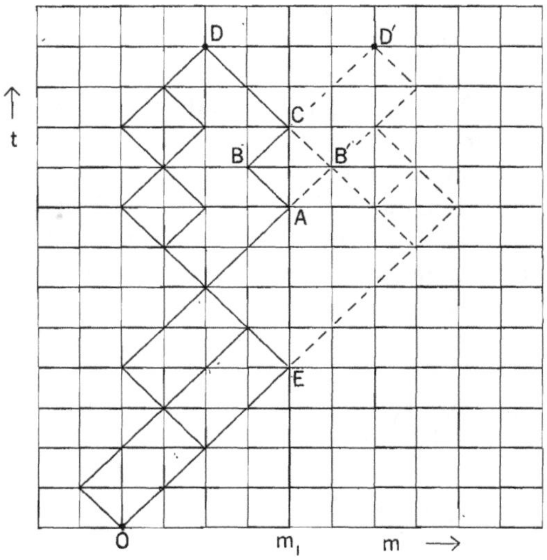

FIG. 1.

one unit while at the same time it suffers a lateral displacement also by one unit either in the positive or in the negative direction.

In the absence of a reflecting wall at $m=m_1$ the probability that the particle arrives at m after N steps is of course given by Eq. (3). But the presence of the reflecting wall requires $W(m, N)$ according to (3) to be modified to take account of the fact that a path reaching m after n reflections must be counted 2^n times since at each reflection it has a probability unity of retracing its step. It is now seen that we can take account of the relevant factors by adding to $W(m, N)$ the probability $W(2m_1-m, N)$ of arriving at the "*image*" point $(2m_1-m)$ after N steps (also in the absence of the reflecting wall), i.e.,

$$W(m, N; m_1) = W(m, N) + W(2m_1-m, N). \quad (18)$$

We can verify the truth of this assertion in the following manner: Consider first a path like OED which has suffered just one reflection at m_1. By reflecting this path about the vertical line through m_1 we obtain a trajectory leading to the image point $(2m_1-m)$ and conversely, for every trajectory leading to the image point, having crossed the line through m_1 once, there is exactly one which leads to m after a single reflection. Thus, instead of counting twice each trajectory reflected once, we can add a uniquely defined trajectory leading to $(2m_1-m)$. Consider next a trajectory like $OABCD$ which leads to m after two reflections. A trajectory like this should be counted four times. But there are two trajectories ($OAB'CD$ and $OABCD'$) leading to the image point and a third ($OAB'CD$) which we should exclude on account of the barrier. These three additional trajectories together with $OABCD$ give exactly four trajectories leading either to m or its image $2m_1-m$ in the absence of the reflecting barrier. In this manner the arguments can be extended to prove the general validity of (18).

If we pass to the limit of large N Eq. (18) becomes [cf. Eq. (12)]

$$W(m,\, N;\, m_1)=\left(\frac{2}{\pi N}\right)^{\frac{1}{2}}\{\exp\,(-m^2/2N)$$
$$+\exp\,[-(2m_1-m)^2/2N]\}. \quad (19)$$

Again, if as in §1 we use the net displacement $x=ml$ as the variable and consider the probability $W(x,\, t;\, x_1)\Delta x$ that the particle is between x and $x+\Delta x$, $(\Delta x\gg l)$ after a time t (during which time it has taken nt steps) in the presence of a reflecting barrier at $x_1=m_1l$, we have

$$W(x,\, t;\, x_1)=\frac{1}{2(\pi Dt)^{\frac{1}{2}}}\{\exp\,(-x^2/4Dt)$$
$$+\exp\,[-(2x_1-x)^2/4Dt]\}. \quad (20)$$

We may note here for future reference that according to Eq. (20)

$$(\partial W/\partial x)_{x=x_1}\equiv 0. \quad (21)$$

(b) *Absorbing Wall at* $m=m_1$

We shall now consider the case when there is a perfectly absorbing barrier at $m=m_1$. The interposition of the perfect absorber at m_1 means that whenever the particle arrives at m_1 it at once becomes incapable of suffering further displacements.[4] There are two questions which we should like to answer under these circumstances. The first is the analog of the problems we have considered so far, namely the probability that the particle arrives at $m(\leqslant m_1)$ after taking N steps. The second question which is characteristic of the present problem concerns the average rate at which the particle will deposit itself on the absorbing screen.

Considering first the probability $W(m,\, N;\, m_1)$, it is clear that in counting the number of distinct sequences of steps which lead to m we should be careful to exclude all sequences which include even a single arrival to m_1. In other words, if we first count *all* possible sequences which lead to m in the absence of the absorbing screen we should then exclude a certain number of "*forbidden*" sequences. It is evident, on the other hand, that every such forbidden sequence uniquely defines another sequence leading to the image $(2m_1-m)$ of m on the line $m=m_1$ in the $(m,\, N)$-plane (see Fig. 1) and conversely. For, by reflecting about the line $m=m_1$ the part of a forbidden trajectory *above* its last point of contact with the line $m=m_1$ before arriving at m we are led to a trajectory leading to the image point, and conversely for every trajectory leading to $2m_1-m$ we necessarily obtain by reflection a forbidden trajectory leading to m (since any trajectory leading to $2m_1-m$ must necessarily cross the line $m=m_1$). Hence,

$$W(m,\, N;\, m_1)=W(m,\, N)-W(2m_1-m,\, N). \quad (22)$$

For large N we have

$$W(m,\, N;\, m_1)=(2/\pi N)^{\frac{1}{2}}\{\exp\,(-m^2/2N)$$
$$-\exp\,[-(2m_1-m)^2/2N]\}. \quad (23)$$

Similarly, analogous to Eq. (21) we now have

$$W(x,\, t;\, x_1)=\frac{1}{2(\pi Dt)^{\frac{1}{2}}}\{\exp\,(-x^2/4Dt)$$
$$-\exp\,[-(2x_1-x)^2/4Dt]\}. \quad (24)$$

We may further note that according to this equation

$$W(x_1,\, t;\, x_1)\equiv 0. \quad (25)$$

Turning next to our second question concerning the probable rate at which the particle deposits itself on the absorbing screen, we may first formulate the problem more specifically. What we wish to know is simply the probability $a(m_1,\, N)$ that after taking N steps the particle will arrive at m_1 *without ever* having touched or crossed the line $m=m_1$ at any earlier step.

First of all it is clear that N should have to be even or odd depending on whether m_1 is even

[4] This problem has important applications to other physical problems.

or odd. We shall suppose that this is the case. Suppose now that there is no absorbing screen. Then the arrival of the particle at m_1 after N steps implies that its position after $(N-1)$ steps must have been either (m_1-1) or (m_1+1). (See Fig. 2.) But every trajectory which arrives at (m_1, N) from $(m_1+1, N-1)$ is a forbidden one in the presence of the absorbing screen since such a trajectory must necessarily have crossed the line $m=m_1$. It does *not* however follow that *all* trajectories arriving at (m_1, N) from $(m_1-1, N-1)$ are permitted ones: For, a certain number of these trajectories will have touched or crossed the line $m=m_1$ earlier than its last step. The number of such trajectories arriving at $(m_1-1, N-1)$ but having an earlier contact with, or a crossing of, the line $m=m_1$ is equal to those arriving at $(m_1+1, N-1)$. The argument is simply that by reflection about the line $m=m_1$ we can uniquely derive from a trajectory leading to $(m_1+1, N-1)$ another leading to $(m_1-1, N-1)$ which has a forbidden character, and conversely. Thus, the number of permitted ways of arriving at m_1 for the first time after N steps is equal to *all* the possible ways of arriving at m_1 after N steps in the absence of the absorbing wall *minus* twice the number of ways of arriving at $(m_1+1, N-1)$ again in the absence of the absorbing screen: i.e.,

$$\frac{N!}{[\frac{1}{2}(N-m_1)]![\frac{1}{2}(N+m_1)]!} - 2\frac{(N-1)!}{[\frac{1}{2}(N+m_1)]![\frac{1}{2}(N-m_1-2)]!}$$
$$=\frac{N!}{[\frac{1}{2}(N-m_1)]![\frac{1}{2}(N+m_1)]!}\left(1-\frac{N-m_1}{N}\right), \quad (26)$$
$$=\frac{m_1}{N}\frac{N!}{[\frac{1}{2}(N-m_1)]![\frac{1}{2}(N+m_1)]!}.$$

The required probability $a(m_1, N)$ is therefore given by

$$a(m_1, N)=\frac{m_1}{N}W(m_1, N). \quad (27)$$

For the limiting case of large N we have

$$a(m_1, N)=\frac{m_1}{N}\left(\frac{2}{\pi N}\right)^{\frac{1}{2}} \exp(-m_1^2/2N). \quad (28)$$

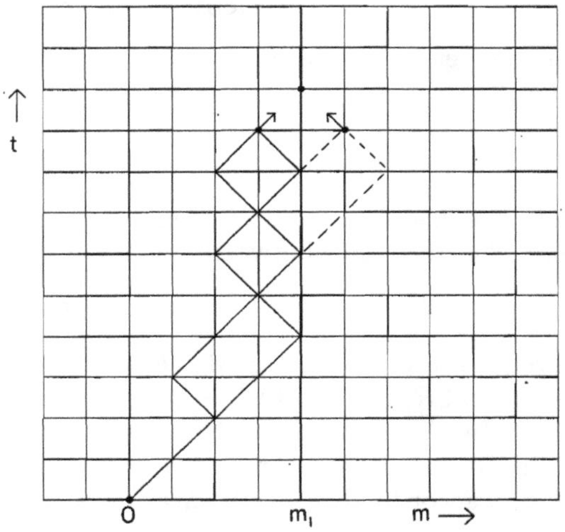

FIG. 2.

If we further write

$$x_1=m_1 l; \quad N=nt; \quad D=\tfrac{1}{2}nl^2, \quad (29)$$

where l is the length of each step and n the number of displacements (assumed constant) which the particle suffers in unit time, then

$$a(x_1, t)=\frac{x_1}{nt}\frac{1}{(\pi Dt)^{\frac{1}{2}}} \exp(-x_1^2/4Dt). \quad (30)$$

Finally, if we ask the probability $q(x_1, t)\Delta t$ that the particle arrives at x_1 during t and $t+\Delta t$ for the first time, then

$$q(x_1, t)\Delta t=\tfrac{1}{2}a(x_1, t)n\Delta t, \quad (31)$$

since (30) is the number which arrive at x_1 in the time taken to traverse two steps. Thus,

$$q(x_1, t)=\frac{x_1}{t}\frac{1}{2(\pi Dt)^{\frac{1}{2}}} \exp(-x_1^2/4Dt). \quad (32)$$

We can interpret Eq. (32) as giving the fraction of a large number of particles initially at $x=0$ and which are deposited on the absorbing screen per unit time, at time t.

We readily verify that $q(x_1, t)$ as defined by Eq. (32) satisfies the relation

$$q(x_1, t)=-D(\partial W/\partial x)_{x=x_1}, \quad (33)$$

with W defined as in Eq. (24). This equation has an important physical interpretation to which we shall draw attention in §5.

3. The General Problem of Random Flights: Markoff's Method

In the general problem of random flights, the position $\boldsymbol{R}$ of the particle after N displacements is given by

$$\boldsymbol{R}=\sum_{i=1}^{N} \boldsymbol{r}_i, \tag{34}$$

where the $\boldsymbol{r}_i$'s $(i=1, \cdots, N)$ denote the different displacements. Further, the probability that the ith displacement lies between $\boldsymbol{r}_i$ and $\boldsymbol{r}_i+d\boldsymbol{r}_i$ is given by

$$\tau_i(x_i, y_i, z_i)dx_idy_idz_i=\tau_id\boldsymbol{r}_i \quad (i=1, \cdots, N). \tag{35}$$

We require the probability $W_N(\boldsymbol{R})d\boldsymbol{R}$ that the position of the particle after N displacements lies in the interval $\boldsymbol{R}$, $\boldsymbol{R}+d\boldsymbol{R}$. In this general form the problem can be solved by using a method originally devised by A. A. Markoff. Now, Markoff's method is of such extreme generality that it actually enables us to solve the first of the two fundamental problems outlined in the introductory section. We shall accordingly describe Markoff's method in a form in which it can readily be applied to other problems besides that of random flights.

Let

$$\boldsymbol{\phi}_j=(\phi_j^1, \phi_j^2, \cdots, \phi_j^n) \quad (j=1, \cdots, N) \tag{36}$$

be N, n-dimensional vectors, the components of each of these vectors being functions of s coordinates:

$$\phi_j^k=\phi_j^k(q_j^1, \cdots, q_j^s) \quad (k=1, \cdots, n;\ j=1, \cdots, N). \tag{37}$$

The probability that the q_j^i's occur in the range

$$q_j^1, q_j^1+dq_j^1;\ q_j^2, q_j^2+dq_j^2;\ \cdots;\ q_j^s, q_j^s+dq_j^s, \quad (j=1, \cdots, N) \tag{38}$$

is given by

$$\tau_j(q_j^1, \cdots, q_j^s)dq_j^1\cdots dq_j^s=\tau_j(\boldsymbol{q}_j)d\boldsymbol{q}_j. \tag{39}$$

Further, let

$$(\Phi^1, \Phi^2, \cdots, \Phi^n)=\boldsymbol{\Phi}=\sum_{j=1}^{N} \boldsymbol{\phi}_j. \tag{40}$$

The problem is: What is the probability that

$$\boldsymbol{\Phi}_0-\tfrac{1}{2}d\boldsymbol{\Phi}_0\leqslant\boldsymbol{\Phi}\leqslant\boldsymbol{\Phi}_0+\tfrac{1}{2}d\boldsymbol{\Phi}_0 \tag{41}$$

where $\boldsymbol{\Phi}_0$ is some preassigned value for $\boldsymbol{\Phi}$.

If we denote the required probability by

$$W_N(\boldsymbol{\Phi}_0)d\Phi_0^1\cdots d\Phi_0^n=W(\boldsymbol{\Phi}_0)d\boldsymbol{\Phi}_0, \tag{42}$$

we clearly have

$$W_N(\boldsymbol{\Phi}_0)d\boldsymbol{\Phi}_0=\int\cdots\int \prod_{j=1}^{N} \{\tau_j(\boldsymbol{q}_j)d\boldsymbol{q}_j\}, \tag{43}$$

where the integration is effected over only those parts of the Ns-dimensional configuration space $(q_1^1, \cdots, q_N^s)$ in which the inequalities (41) are satisfied.

We shall now introduce a factor $\Delta(\boldsymbol{q}_1, \cdots, \boldsymbol{q}_N)$ having the following properties:

$$\left.\begin{aligned}\Delta(\boldsymbol{q}_1, \cdots, \boldsymbol{q}_N)&=1 \quad \text{whenever} \quad \boldsymbol{\Phi}_0-\tfrac{1}{2}d\boldsymbol{\Phi}_0\leqslant\boldsymbol{\Phi}\leqslant\boldsymbol{\Phi}_0+\tfrac{1}{2}d\boldsymbol{\Phi}_0,\\ &=0 \quad \text{otherwise.}\end{aligned}\right\} \tag{44}$$

Then,

$$W_N(\boldsymbol{\Phi}_0)d\boldsymbol{\Phi}_0=\int\cdots\int \Delta(\boldsymbol{q}_1, \cdots, \boldsymbol{q}_N) \prod_{j=1}^{N} \{\tau_j(\boldsymbol{q}_j)d\boldsymbol{q}_j\} \tag{45}$$

where, in contrast to (43), the integration is now extended over *all* the accessible regions of the configuration space. The introduction of the factor Δ under the integral sign in Eq. (45) in this manner appears at first sight as a very formal device to extend the range of integration over the entire configuration space. But the essence of Markoff's method is that an explicit expression for this factor can be given.

Consider the integrals

$$\delta_k = \frac{1}{\pi}\int_{-\infty}^{+\infty} \frac{\sin \alpha_k \rho_k}{\rho_k} \exp(i\rho_k \gamma_k) d\rho_k \quad (k=1, \cdots, n). \tag{46}$$

The integral defining δ_k is the well-known discontinuous integral of Dirichlet and has the property

$$\left.\begin{aligned} \delta_k &= 1 \quad \text{whenever} \quad -\alpha_k < \gamma_k < \alpha_k, \\ &= 0 \quad \text{otherwise.} \end{aligned}\right\} \tag{47}$$

Now, let

$$\alpha_k = \tfrac{1}{2} d\Phi_0{}^k; \quad \gamma_k = \sum_{j=1}^{N} \phi_j{}^k - \Phi_0{}^k \quad (k=1, \cdots, n). \tag{48}$$

According to Eq. (47)

$$\left.\begin{aligned} \delta_k &= 1 \quad \text{whenever} \quad \Phi_0{}^k - \tfrac{1}{2} d\Phi_0{}^k < \sum_{j=1}^{N} \phi_j{}^k < \Phi_0{}^k + \tfrac{1}{2} d\Phi_0{}^k, \\ &= 0 \quad \text{otherwise.} \end{aligned}\right\} \tag{49}$$

Consequently

$$\Delta = \prod_{k=1}^{n} \delta_k \tag{50}$$

has the required properties (44).

Substituting for Δ from Eqs. (46) and (50) in Eq. (45), we obtain

$$\left.\begin{aligned} W_N(\mathbf{\Phi}_0) d\mathbf{\Phi}_0 &= \frac{1}{\pi^n} \int \underset{(\mathbf{\varrho})}{\cdots} \int \int \underset{(\mathbf{q})}{\cdots} \int \left\{ \prod_{j=1}^{N} \tau_j(\mathbf{q}_j) d\mathbf{q}_j \right\} \left\{ \prod_{k=1}^{n} \frac{\sin(\frac{1}{2} d\Phi_0{}^k \rho_k)}{\rho_k} \right\} \\ &\quad \times \exp\left\{ i \left[\sum_{k=1}^{n} \sum_{j=1}^{N} \phi_j{}^k \rho_k - \sum_{k=1}^{n} \Phi_0{}^k \rho_k \right] \right\} d\rho_1 \cdots d\rho_n \\ &= \frac{d\mathbf{\Phi}_0}{2^n \pi^n} \int \cdots \int \exp(-i\mathbf{\varrho} \cdot \mathbf{\Phi}_0) A_N(\mathbf{\varrho}) d\mathbf{\varrho} \end{aligned}\right\} \tag{51}$$

where we have written

$$A_N(\mathbf{\varrho}) = \prod_{j=1}^{N} \int \cdots \int dq_j{}^1 \cdots dq_j{}^s \exp(i\mathbf{\varrho} \cdot \mathbf{\phi}_j) \tau_j(q_j{}^1, \cdots, q_j{}^s). \tag{52}$$

The case of greatest interest is when all the functions τ_j (of the respective q_j's) are identical. Equation (52) then becomes

$$A_N(\mathbf{\varrho}) = \left[\int \exp(i\mathbf{\varrho} \cdot \mathbf{\phi}) \tau(\mathbf{q}) d\mathbf{q} \right]^N. \tag{53}$$

According to Eq. (51), $A_N(\mathbf{\varrho})$ is the n-dimensional Fourier-transform of the probability function $W(\mathbf{\Phi}_0)$. And Markoff's procedure illustrates a very general principle that it is the Fourier transform of the probability function, rather than the function itself, that has a more direct relation to the physical situations.

For $N\rightarrow\infty$, $A_N(\boldsymbol{\rho})$ generally tends to the form [see §4 Eq. (91)]

$$\underset{N\rightarrow\infty}{\text{Limit}}\, A_N(\boldsymbol{\rho})=\exp\left[-C(\boldsymbol{\rho})\right]. \tag{54}$$

4. The Solution to the General Problem of Random Flights

We shall now apply Markoff's method to the problem of random flights. According to Eqs. (34), (51), and (52), the probability $W_N(\boldsymbol{R})d\boldsymbol{R}$ that the position $\boldsymbol{R}$ of the particle will be found in the interval $(\boldsymbol{R}, \boldsymbol{R}+d\boldsymbol{R})$ after N displacements is given by

$$W_N(\boldsymbol{R})=\frac{1}{8\pi^3}\int_{-\infty}^{+\infty}\exp\left(-i\boldsymbol{\rho}\cdot\boldsymbol{R}\right)A_N(\boldsymbol{\rho})d\boldsymbol{\rho}, \tag{55}$$

where

$$A_N(\boldsymbol{\rho})=\prod_{j=1}^{N}\int_{-\infty}^{+\infty}\tau_j(\boldsymbol{r}_j)\exp\left(i\boldsymbol{\rho}\cdot\boldsymbol{r}_j\right)d\boldsymbol{r}_j. \tag{56}$$

In Eq. (55), $\tau_j(\boldsymbol{r}_j)$ governs the probability of occurrence of a displacement $\boldsymbol{r}_j$ on the jth occasion. The explicit form which $W_N(\boldsymbol{R})$ takes will naturally depend on the assumptions made concerning the $\tau_j(\boldsymbol{r}_j)$'s. We shall now consider several cases of interest.

(a) *A Gaussian Distribution of the Displacements* $\boldsymbol{r}_j$

A case of special interest arises when

$$\tau_j=\frac{1}{(2\pi l_j^2/3)^{\frac{3}{2}}}\exp\left(-3|\boldsymbol{r}_j|^2/2l_j^2\right), \tag{57}$$

where l_j^2 denotes the mean square displacement to be expected on the jth occasion. While l_j^2 may differ from one displacement to another we assume that *all* the displacements occur in random directions.

For τ_j of the form (57), our expression for $A_N(\boldsymbol{\rho})$ becomes

$$\begin{aligned}A_N(\boldsymbol{\rho})&=\prod_{j=1}^{N}\frac{1}{(2\pi l_j^2/3)^{\frac{3}{2}}}\iiint_{-\infty}^{+\infty}\exp\left[i(\rho_1x_j+\rho_2y_j+\rho_3z_j)-3(x_j^2+y_j^2+z_j^2)/2l_j^2\right]dx_jdy_jdz_j\\&=\prod_{j=1}^{N}\exp\left[-(\rho_1^2+\rho_2^2+\rho_3^2)l_j^2/6\right]=\exp\left[-(|\boldsymbol{\rho}|^2\sum_{j=1}^{N}l_j^2)/6\right].\end{aligned} \tag{58}$$

Let $\langle l^2\rangle_{\text{Av}}$ stand for

$$\langle l^2\rangle_{\text{Av}}=\frac{1}{N}\sum_{j=1}^{N}l_j^2. \tag{59}$$

Equation (58) becomes

$$A_N(\boldsymbol{\rho})=\exp\left[-N\langle l^2\rangle_{\text{Av}}|\boldsymbol{\rho}|^2/6\right]. \tag{60}$$

Substituting this expression for $A_N(\boldsymbol{\rho})$ in Eq. (55), we obtain

$$W_N(\boldsymbol{R})=\frac{1}{8\pi^3}\iiint_{-\infty}^{+\infty}\exp\left[-i(\rho_1X+\rho_2Y+\rho_3Z)-N\langle l^2\rangle_{\text{Av}}(\rho_1^2+\rho_2^2+\rho_3^2)/6\right]d\rho_1d\rho_2d\rho_3. \tag{61}$$

The integrations in (61) are readily performed and we find

$$W_N(\boldsymbol{R})=\frac{1}{(2\pi N\langle l^2\rangle_{\text{Av}}/3)^{\frac{3}{2}}}\exp\left[-3|\boldsymbol{R}|^2/2N\langle l^2\rangle_{\text{Av}}\right]. \tag{62}$$

This is an *exact* solution valid for any value of N. That an exact solution can be found for a Gaussian distribution of the different displacements is simply a consequence of the "*addition theorem*" which these functions satisfy.

(b) Each Displacement of a Constant Length But in Random Directions

Let the displacement on the jth occasion be of length l_j in a random direction. Under these circumstances, we can define the distribution functions τ_j by

$$\tau_j = \frac{1}{4\pi l_j^3}\delta(|\boldsymbol{r}_j|^2 - l_j^2), \quad (j=1, \cdots, N) \tag{63}$$

where δ stands for Dirac's δ function.

Accordingly, our expression for $A_N(\boldsymbol{\varrho})$ becomes

$$A_N(\boldsymbol{\varrho}) = \prod_{j=1}^{N} \frac{1}{4\pi l_j^3} \int_{-\infty}^{+\infty} \exp\,(i\boldsymbol{\varrho}\cdot\boldsymbol{r}_j)\,\delta(r_j^2 - l_j^2)\,d\boldsymbol{r}_j, \tag{64}$$

or, using polar coordinates with the z axis in the direction of $\boldsymbol{\varrho}$

$$A_N(\boldsymbol{\varrho}) = \prod_{j=1}^{N} \frac{1}{4\pi l_j^3} \int_0^\infty \int_0^\pi \int_0^{2\pi} \exp\,[i|\boldsymbol{\varrho}|r_j\cos\vartheta]\,\delta(r_j^2 - l_j^2)\,r_j^2 \sin\vartheta\, dr_j d\vartheta d\omega. \tag{65}$$

The integrations over the polar and the azimuthal angles ϑ and ω are readily effected:

$$\left.\begin{aligned} A_N(\boldsymbol{\varrho}) &= \prod_{j=1}^{N} \frac{1}{2l_j^3} \int_0^\infty \int_0^\pi \exp\,(i|\boldsymbol{\varrho}|r_j\cos\vartheta)\,r_j^2\delta(r_j^2 - l_j^2)\sin\vartheta\, d\vartheta dr_j \\ &= \prod_{j=1}^{N} \frac{1}{l_j^3|\boldsymbol{\varrho}|} \int_0^\infty \sin\,(|\boldsymbol{\varrho}|r_j)\,r_j\delta(r_j^2 - l_j^2)\,dr_j \\ &= \prod_{j=1}^{N} \frac{\sin\,(|\boldsymbol{\varrho}|l_j)}{|\boldsymbol{\varrho}|l_j}. \end{aligned}\right\} \tag{66}$$

Thus,

$$W_N(\boldsymbol{R}) = \frac{1}{8\pi^3} \int_{-\infty}^{+\infty} \exp\,(-i\boldsymbol{\varrho}\cdot\boldsymbol{R}) \prod_{j=1}^{N} \frac{\sin\,(|\boldsymbol{\varrho}|l_j)}{|\boldsymbol{\varrho}|l_j}\,d\boldsymbol{\varrho}. \tag{67}$$

Again, choosing polar coordinates but with the z axis pointing this time in the direction of $\boldsymbol{R}$, we have

$$W_N(\boldsymbol{R}) = \frac{1}{8\pi^3} \int_0^\infty \int_{-1}^{+1} \int_0^{2\pi} \exp\,(-i|\boldsymbol{\varrho}|\,|\boldsymbol{R}|t) \left\{ \prod_{j=1}^{N} \frac{\sin\,(|\boldsymbol{\varrho}|l_j)}{|\boldsymbol{\varrho}|l_j} \right\} |\boldsymbol{\varrho}|^2 d\omega dt d|\boldsymbol{\varrho}|. \tag{68}$$

The integrations over ω and t are readily performed and we obtain

$$W_N(\boldsymbol{R}) = \frac{1}{2\pi^2|\boldsymbol{R}|} \int_0^\infty \sin\,(|\boldsymbol{\varrho}|\,|\boldsymbol{R}|) \left\{ \prod_{j=1}^{N} \frac{\sin\,(|\boldsymbol{\varrho}|l_j)}{|\boldsymbol{\varrho}|l_j} \right\} |\boldsymbol{\varrho}|\,d|\boldsymbol{\varrho}| \tag{69}$$

which represents the formal solution to the problem. In this form, the solution for the problem of random flights is due to Rayleigh.[5]

---

[5] Lord Rayleigh, *Collected Papers*, Vol. 6, p. 604. We may, however, draw attention to the fact that our formulation of the general problem of random flights is wider in its scope than Rayleigh's. Rayleigh's formulation of the problem corresponds to our special case (63).

The case of greatest interest arises when all the l_j's are equal. We shall assume that this is the case in the rest of our discussion:

$$l_j = l = \text{constant} \quad (j=1, \cdots, N). \tag{70}$$

Equation (69) becomes

$$W_N(R) = \frac{1}{2\pi^2 |R|} \int_0^\infty \sin(|\varrho||R|) \left\{ \frac{\sin(|\varrho| l)}{|\varrho| l} \right\}^N |\varrho| d|\varrho|. \tag{71}$$

(*i*) *N finite.*—We shall illustrate (following Rayleigh) the method of evaluating the integral on the right-hand side of Eq. (71) for finite values of N by considering the cases $N=3$ and 4.

When $N=3$, Eq. (71) becomes

$$W_3(R) = \frac{1}{2\pi^2 |R| l^3} \int_0^\infty \sin(|\varrho||R|) \sin^3(|\varrho| l) \frac{d|\varrho|}{|\varrho|^2}. \tag{72}$$

But

$$\sin(|\varrho||R|)\sin^3(|\varrho| l) = \tfrac{1}{8}\{3\cos[(|R|-l)|\varrho|] - 3\cos[(|R|+l)|\varrho|] - \cos[(|R|-3l)|\varrho|] + \cos[(|R|+3l)|\varrho|]\}. \tag{73}$$

Further

$$\left.\begin{aligned} \int_0^\infty \{\cos[(|R|-l)|\varrho|] - \cos[(|R|+l)|\varrho|]\} \frac{d|\varrho|}{|\varrho|^2} \\ = 2\int_0^\infty \left\{ \sin^2 \frac{(|R|+l)|\varrho|}{2} - \sin^2 \frac{(|R|-l)|\varrho|}{2} \right\} \frac{d|\varrho|}{|\varrho|^2} \\ = \tfrac{1}{2}\pi(|R|+l-||R|-l|). \end{aligned}\right\} \tag{74}$$

We have a similar formula for the integral involving the other pair of cosines in Eq. (73). Combining these results we obtain

$$W_3(R) = \frac{1}{32\pi |R| l^3}\{2|R| - 3||R|-l| + ||R|-3l|\}, \tag{75}$$

or, more explicitly

$$\left.\begin{aligned} W_3(R) &= \frac{1}{8\pi l^3} && (0<|R|<l), \\ &= \frac{1}{16\pi l^3 |R|}(3l-|R|) && (l<|R|<3l), \\ &= 0 && (3l<|R|<\infty). \end{aligned}\right\} \tag{76}$$

We shall consider next the case $N=4$. According to Eq. (71) we have

$$W_4(R) = \frac{1}{2\pi^2 |R| l^4} \int_0^\infty \frac{d|\varrho|}{|\varrho|^3} \sin(|\varrho||R|) \sin^4(|\varrho| l). \tag{77}$$

From this equation we derive

$$
\begin{aligned}
-\frac{d^2}{d|\boldsymbol{R}|^2}[|\boldsymbol{R}|\,W_4(\boldsymbol{R})] &= \frac{1}{2\pi^2 l^4}\int_0^\infty \frac{d|\varrho|}{|\varrho|}\sin(|\varrho|\,|\boldsymbol{R}|)\sin^4(|\varrho|l) \\
&= \frac{1}{32\pi^2 l^4}\int_0^\infty \frac{d|\varrho|}{|\varrho|}\{\sin[(|\boldsymbol{R}|+4l)|\varrho|]+\sin[(|\boldsymbol{R}|-4l)|\varrho|] \\
&\qquad -4\sin[(|\boldsymbol{R}|+2l)|\varrho|]-4\sin[(|\boldsymbol{R}|-2l)|\varrho|]+6\sin(|\boldsymbol{R}|\,|\varrho|)\} \\
&= \frac{1}{64\pi l^4}(1\pm 1-4\mp 4+6)=\frac{1}{64\pi l^4}(3\pm 1\mp 4),
\end{aligned} \tag{78}
$$

where the two alternatives in the last two steps of Eq. (78) depend, respectively, on the signs of $(|\boldsymbol{R}|-4l)$ and $(|\boldsymbol{R}|-2l)$. Thus

$$
\begin{aligned}
64\pi l^4\frac{d^2}{d|\boldsymbol{R}|^2}[|\boldsymbol{R}|\,W_4(\boldsymbol{R})] &= -6 \quad (0<|\boldsymbol{R}|<2l), \\
&= +2 \quad (2l<|\boldsymbol{R}|<4l), \\
&= 0 \quad (4l<|\boldsymbol{R}|<\infty).
\end{aligned} \tag{79}
$$

We can integrate the foregoing equation working backwards from large values of $|\boldsymbol{R}|$ where all derivatives must vanish. We find

$$
\begin{aligned}
64\pi l^4\frac{d}{d|\boldsymbol{R}|}[|\boldsymbol{R}|\,W_4(\boldsymbol{R})] &= 2(|\boldsymbol{R}|-4l) \quad (2l<|\boldsymbol{R}|<4l), \\
&= -6|\boldsymbol{R}|+8l \quad (0<|\boldsymbol{R}|<2l),
\end{aligned} \tag{80}
$$

where we have used the continuity of the quantity on the left-hand side of this equation at $|\boldsymbol{R}|=2l$. Integrating Eq. (80) once again we similarly obtain

$$
\left.\begin{aligned}
64\pi l^4|\boldsymbol{R}|\,W_4(\boldsymbol{R}) &= |\boldsymbol{R}|^2-8l|\boldsymbol{R}|+16l^2 \\
&= (4l-|\boldsymbol{R}|)^2
\end{aligned}\right\}(2l<|\boldsymbol{R}|<4l), \tag{81}
$$

and

$$
64\pi l^4|\boldsymbol{R}|\,W_4(\boldsymbol{R}) = -3|\boldsymbol{R}|^2+8l|\boldsymbol{R}|\,(2l>|\boldsymbol{R}|>0). \tag{82}
$$

Thus, finally

$$
\begin{aligned}
W_4(\boldsymbol{R}) &= \frac{1}{64\pi l^4|\boldsymbol{R}|}(8l|\boldsymbol{R}|-3|\boldsymbol{R}|^2) \quad (0<|\boldsymbol{R}|<2l), \\
&= \frac{1}{64\pi l^4|\boldsymbol{R}|}(4l-|\boldsymbol{R}|)^2 \quad (2l<|\boldsymbol{R}|<4l), \\
&= 0 \quad (4l<|\boldsymbol{R}|<\infty).
\end{aligned} \tag{83}
$$

In like manner it is possible, in principle, to evaluate the integral for $W_N(\boldsymbol{R})$ for any finite value of N. But the calculations become very tedious. We may however note the following solution obtained by Rayleigh for the case $N=6$.

$$\left.\begin{aligned} W_6(\boldsymbol{R}) &= \frac{1}{2^8\pi|\boldsymbol{R}|l^6}(16l^3|\boldsymbol{R}|-4l|\boldsymbol{R}|^3+(5/6)|\boldsymbol{R}|^4) && (0<R<2l) \\ &= \frac{1}{2^8\pi|\boldsymbol{R}|l^6}(-20l^4+56l^3|\boldsymbol{R}|-30l^2|\boldsymbol{R}|^2+6l|\boldsymbol{R}|^3-(5/12)|\boldsymbol{R}|^4) && (2l<|\boldsymbol{R}|<4l) \\ &= \frac{1}{2^8\pi|\boldsymbol{R}|l^6}(108l^4-72l^3|\boldsymbol{R}|+18l^2|\boldsymbol{R}|^2-2l|\boldsymbol{R}|^3+(1/12)|\boldsymbol{R}|^4) && (4l<|\boldsymbol{R}|<6l) \\ &= 0 && (6l<|\boldsymbol{R}|<\infty). \end{aligned}\right\} \tag{84}$$

(*ii*) $N \ll 1$.—By far the most interesting case is when N is very large. Under these circumstances

$$\left.\begin{aligned} \underset{N\to\infty}{\text{Limit}}\left(\frac{\sin(|\varrho|l)}{|\varrho|l}\right)^N &= \underset{N\to\infty}{\text{Limit}}\,(1-\tfrac{1}{6}|\varrho|^2l^2+\cdots)^N, \\ &= \exp(-N|\varrho|^2l^2/6). \end{aligned}\right\} \tag{85}$$

Accordingly, from Eq. (69) we conclude that for large values of N

$$W(\boldsymbol{R}) = \frac{1}{2\pi^2|\boldsymbol{R}|}\int_0^\infty \exp(-Nl^2|\varrho|^2/6)|\varrho|\sin(|\boldsymbol{R}||\varrho|)d|\varrho|, \tag{86}$$

where we have written $W(\boldsymbol{R})$ for $W_N(\boldsymbol{R})$, $N\to\infty$. Evaluating the integral on the right-hand side of Eq. (86), we find

$$W(\boldsymbol{R}) = \frac{1}{(2\pi Nl^2/3)^{\frac{3}{2}}}\exp(-3|\boldsymbol{R}|^2/2Nl^2). \tag{87}$$

We notice the formal similarity of Eqs. (62) and (87). However, on our present assumptions, Eq. (87) is valid only for large values of N.

(*c*) *A Spherical Distribution of the Displacements.* $N \gg 1$

We shall assume that

$$\tau_j(\boldsymbol{r}_j) = \tau(|\boldsymbol{r}_j|^2) \quad (j=1,\cdots,N). \tag{88}$$

Then

$$A_N(\varrho) = \left[\int_{-\infty}^{+\infty}\exp(i\varrho\cdot\boldsymbol{r})\tau(r^2)d\boldsymbol{r}\right]^N. \tag{89}$$

By using polar coordinates, the integral inside the square brackets in Eq. (89) becomes

$$\int_{-\infty}^{+\infty}\exp(i\varrho\cdot\boldsymbol{r})\tau(r^2)d\boldsymbol{r} = \int_0^\infty\int_{-1}^{+1}\int_0^{2\pi}\exp(i|\varrho|rt)r^2\tau(r^2)d\omega dt dr = 4\pi\int_0^\infty\frac{\sin(|\varrho|r)}{|\varrho|r}r^2\tau(r^2)dr. \tag{90}$$

Hence

$$\left.\begin{aligned} \underset{N\to\infty}{\text{Limit}}\,A_N(\varrho) &= \underset{N\to\infty}{\text{Limit}}\left[4\pi\int_0^\infty\frac{\sin(|\varrho|r)}{|\varrho|r}r^2\tau(r^2)dr\right]^N, \\ &= \underset{N\to\infty}{\text{Limit}}\left[4\pi\int_0^\infty(1-\tfrac{1}{6}|\varrho|^2r^2+\cdots)r^2\tau(r^2)dr\right]^N, \\ &= \exp(-N|\varrho|^2\langle r^2\rangle_{\text{Av}}/6) \end{aligned}\right\} \tag{91}$$

where $\langle r^2\rangle_{\text{Av}}$ is the mean square displacement to be expected on any occasion. Substituting the foregoing result in Eq. (55) we obtain

$$W(\mathbf{R})=\frac{1}{8\pi^3}\int_{-\infty}^{+\infty}\exp\left(-i\boldsymbol{\rho}\cdot\mathbf{R}-N|\boldsymbol{\rho}|^2\langle r^2\rangle_{\text{Av}}/6\right)d\boldsymbol{\rho}, \tag{92}$$

or, [cf. Eq. (62)]

$$W(\mathbf{R})=\frac{1}{(2\pi N\langle r^2\rangle_{\text{Av}}/3)^{\frac{3}{2}}}\exp\left(-3|\mathbf{R}|^2/2N\langle r^2\rangle_{\text{Av}}\right). \tag{93}$$

It is seen that Eq. (93) includes the result obtained earlier in Section (b) under case (ii) [Eq. (87)] as a special case.

(d) *The Solution to the General Problem of Random Flights for $N\gg 1$*

We shall now obtain the general expression for $W_N(\mathbf{R})$ for large values of N with no special assumptions concerning the distribution of the different displacements except that all the τ_j's represent the same function. Accordingly, we have to examine quite generally the behavior for $N\rightarrow\infty$ of $A_N(\boldsymbol{\rho})$ defined by [cf. Eq. (53)]

$$A_N(\boldsymbol{\rho})=\left[\int_{-\infty}^{+\infty}\exp(i\boldsymbol{\rho}\cdot\mathbf{r})\tau(\mathbf{r})d\mathbf{r}\right]^N. \tag{94}$$

Let ρ_1, ρ_2, ρ_3 denote the components of $\boldsymbol{\rho}$ in some fixed system of coordinates. Then

$$\left.\begin{aligned}
A_N(\boldsymbol{\rho})&=\left[\int_{-\infty}^{+\infty}\int_{-\infty}^{+\infty}\int_{-\infty}^{+\infty}\exp\left[i(\rho_1x+\rho_2y+\rho_3z)\right]\tau(x,y,z)dxdydz\right]^N,\\
&=\left[\int_{-\infty}^{+\infty}\int_{-\infty}^{+\infty}\int_{-\infty}^{+\infty}\{1+i(\rho_1x+\rho_2y+\rho_3z)-\tfrac{1}{2}(\rho_1^2x^2+\rho_2^2y^2+\rho_3^2z^2+2\rho_1\rho_2xy\right.\\
&\qquad\left.+2\rho_2\rho_3yz+2\rho_3\rho_1zx)+\cdots\}\tau(x,y,z)dxdydz\right]^N,\\
&=\left[1+i(\rho_1\langle x\rangle+\rho_2\langle y\rangle+\rho_3\langle z\rangle)-\tfrac{1}{2}(\rho_1^2\langle x^2\rangle+\rho_2^2\langle y^2\rangle+\rho_3^2\langle z^2\rangle+2\rho_1\rho_2\langle xy\rangle\right.\\
&\qquad\left.+2\rho_2\rho_3\langle yz\rangle+2\rho_3\rho_1\langle zx\rangle)+\cdots\right]^N
\end{aligned}\right\} \tag{95}$$

where $\langle x\rangle$, $\cdots$, $\langle zx\rangle$ denote the various first and second moments of the function $\tau(x, y, z)$. Hence for $N\rightarrow\infty$ we have

$$A_N(\boldsymbol{\rho})=\exp\left[iN(\rho_1\langle x\rangle+\rho_2\langle y\rangle+\rho_3\langle z\rangle)-\tfrac{1}{2}NQ(\boldsymbol{\rho})\right] \tag{96}$$

where $Q(\boldsymbol{\rho})$ stands for the homogeneous quadratic form

$$Q(\boldsymbol{\rho})=\langle x^2\rangle\rho_1^2+\langle y^2\rangle\rho_2^2+\langle z^2\rangle\rho_3^2+2\langle xy\rangle\rho_1\rho_2+2\langle yz\rangle\rho_2\rho_3+2\langle zx\rangle\rho_3\rho_1. \tag{97}$$

Substituting for $A_N(\boldsymbol{\rho})$ from Eq. (96) in Eq. (55) we obtain for the probability distribution for large values of N the expression:

$$W(\mathbf{R})=\frac{1}{8\pi^3}\int_{-\infty}^{+\infty}\int_{-\infty}^{+\infty}\int_{-\infty}^{+\infty}\exp\left[-\tfrac{1}{2}NQ(\boldsymbol{\rho})-i\{\rho_1(X-N\langle x\rangle)+\rho_2(Y-N\langle y\rangle)+\rho_3(Z-N\langle z\rangle)\}\right]d\rho_1d\rho_2d\rho_3. \tag{98}$$

To evaluate this integral we first rotate our coordinate system to bring the quadratic form $Q(\boldsymbol{\rho})$ to its diagonal form.

$$Q(\boldsymbol{\rho})=\langle\xi^2\rangle\rho_\xi^2+\langle\eta^2\rangle\rho_\eta^2+\langle\zeta^2\rangle\rho_\zeta^2. \tag{99}$$

In Eq. (99) $\langle\xi^2\rangle$, $\langle\eta^2\rangle$ and $\langle\zeta^2\rangle$ are the eigenvalues of the symmetric matrix formed by the second moments:

$$\begin{vmatrix} \langle x^2\rangle & \langle xy\rangle & \langle xz\rangle \\ \langle yx\rangle & \langle y^2\rangle & \langle yz\rangle \\ \langle zx\rangle & \langle zy\rangle & \langle z^2\rangle \end{vmatrix} \tag{100}$$

Further, the three eigenvectors of the matrix (100) form an orthogonal system which we have denoted by (ξ, η, ζ). Let

$$\boldsymbol{R}=(\Xi, \mathrm{H}, \mathrm{Z}) \tag{101}$$

in this system of coordinates. Equation (98) now reduces to

$$W(\boldsymbol{R})=\frac{1}{8\pi^3}\int_{-\infty}^{+\infty}\int_{-\infty}^{+\infty}\int_{-\infty}^{+\infty}\exp\left[-\tfrac{1}{2}N(\langle\xi^2\rangle\rho_\xi^2+\langle\eta^2\rangle\rho_\eta^2+\langle\zeta^2\rangle\rho_\zeta^2)\right.$$
$$\left.-i\{\rho_\xi(\Xi-N\langle\xi\rangle)+\rho_\eta(\mathrm{H}-N\langle\eta\rangle)+\rho_\zeta(\mathrm{Z}-N\langle\zeta\rangle)\}\right]d\rho_\xi d\rho_\eta d\rho_\zeta. \tag{102}$$

The integrations over ρ_ξ, ρ_η and ρ_ζ are now readily performed, and we find

$$W(\boldsymbol{R})=\frac{1}{(8\pi^3N^3\langle\xi^2\rangle\langle\eta^2\rangle\langle\zeta^2\rangle)^{\frac{1}{2}}}\exp\left[-\frac{(\Xi-N\langle\xi\rangle)^2}{2N\langle\xi^2\rangle}-\frac{(\mathrm{H}-N\langle\eta\rangle)^2}{2N\langle\eta^2\rangle}-\frac{(\mathrm{Z}-N\langle\zeta\rangle)^2}{2N\langle\zeta^2\rangle}\right]. \tag{103}$$

According to Eq. (103), the probability distribution $W(\boldsymbol{R})$ of the position $\boldsymbol{R}$ of the particle after suffering a large number of displacements (governed by a basic distribution function $\tau[x, y, z]$) is an *ellipsoidal distribution* centered at $(N\langle\xi\rangle, N\langle\eta\rangle, N\langle\zeta\rangle)$—in other words the particle suffers an average systematic net displacement of amount $(N\langle\xi\rangle, N\langle\eta\rangle, N\langle\zeta\rangle)$ and superposed on this a general random distribution.

The principal axes of this ellipsoidal distribution are along the principal directions of the moment-ellipsoid defined by (100) and the mean square net displacements about $(N\langle\xi\rangle, N\langle\eta\rangle, N\langle\zeta\rangle)$ along the three principal directions are

$$\langle(\Xi-N\langle\xi\rangle)^2\rangle_{\mathrm{Av}}=N\langle\xi^2\rangle;\quad \langle(\mathrm{H}-N\langle\eta\rangle)^2\rangle_{\mathrm{Av}}=N\langle\eta^2\rangle;\quad \langle(\mathrm{Z}-N\langle\zeta\rangle)^2\rangle_{\mathrm{Av}}=N\langle\zeta^2\rangle. \tag{104}$$

5. The Passage to a Differential Equation: The Reduction of the Problem of Random Flights for Large N to a Boundary Value Problem

In the preceding sections we have obtained the solution to the problem of random flights under various conditions. Though in each case the problem was first formulated and solved for a finite number of displacements, the greatest interest is attached to the limiting form of the solutions for large values of N. And, for large values of N the solutions invariably take very simple forms. Thus, according to Eq. (93) a particle starting from the origin and suffering n displacements per unit time, each displacement $\boldsymbol{r}$ being governed by a probability distribution $\tau(|\boldsymbol{r}|^2)$, will find itself in the element of volume defined by $\boldsymbol{R}$ and $\boldsymbol{R}+d\boldsymbol{R}$ after a time t with the probability

$$W(\boldsymbol{R})d\boldsymbol{R}=\frac{1}{(2\pi n\langle r^2\rangle_{\mathrm{Av}}t/3)^{\frac{3}{2}}}\exp\left(-3|\boldsymbol{R}|^2/2n\langle r^2\rangle_{\mathrm{Av}}t\right)d\boldsymbol{R}. \tag{105}$$

In the foregoing equation $\langle r^2\rangle_{\mathrm{Av}}$ denotes the mean square displacement that is to be expected on any given occasion. If we put

$$D=n\langle r^2\rangle_{\mathrm{Av}}/6 \tag{106}$$

Eq. (105) takes the form [cf. Eq. (16)]

$$W(\boldsymbol{R})d\boldsymbol{R}=\frac{1}{(4\pi Dt)^{\frac{3}{2}}}\exp\left(-|\boldsymbol{R}|^2/4Dt\right)d\boldsymbol{R}. \tag{107}$$

In view of the simplicity of this and the other solutions, the question now arises whether we cannot obtain the asymptotic distributions directly, without passing to the limit of large N, in each case, individually. This problem is of particular importance when restrictions on the motion of the particle in the form of reflecting and absorbing barriers are introduced. Our discussion in §2 of the simple problem of random walk in one dimension with such restrictions already indicates how very complicated the method of enumeration must become under even somewhat more general conditions than those contemplated in §2. The fact, however, that for the solutions obtained in §2, W vanishes on an absorbing wall [Eq. (25)] while grad W vanishes on a reflecting wall [Eq. (21)] suggests that the solutions perhaps correspond to solving a partial differential equation with appropriate boundary conditions. We shall now show how this passage to a differential equation and a boundary value problem is to be achieved.

First, we shall introduce a somewhat different language from that we have used so far in discussing the problem of random flights. Up to the present we have spoken of a *single* particle suffering displacements according to a given probability law, and asking for the probability of finding this particle in some given element of volume at a later time. It is clear that we can instead imagine a very large number of particles starting under the same initial conditions and undergoing the displacements without any mutual interference, and ask the *fraction* of the original number which will be found in a given element of volume at a later time. On this picture, the interpretation of the quantity on the right-hand side of Eq. (106) is that it represents the fraction of a large number of particles which will be found between $\boldsymbol{R}$ and $\boldsymbol{R}+d\boldsymbol{R}$ at time t if all the particles started from $\boldsymbol{R}=0$ at $t=0$. However, the two methods of interpretation are fully equivalent and we shall adopt the language of whichever of the two happens to be more convenient.

We pass on to considerations which lead to a differential equation for $W(\boldsymbol{R}, t)$:

Let Δt denote an interval of time long enough for a particle to suffer a large number of displacements but *still* short enough for the net mean square increment $\langle|\Delta\boldsymbol{R}|^2\rangle_{\text{Av}}$ in $\boldsymbol{R}$ to be small. Under these circumstances, the probability that a particle suffers a net displacement $\Delta\boldsymbol{R}$ in time Δt is given by

$$\psi(\Delta\boldsymbol{R};\Delta t)=\frac{1}{(4\pi D\Delta t)^{\frac{3}{2}}}\exp\left(-|\Delta\boldsymbol{R}|^2/4D\Delta t\right) \tag{108}$$

and is independent of $\boldsymbol{R}$. With Δt chosen in this manner, we seek to derive the probability distribution $W(\boldsymbol{R}, t+\Delta t)$ at time $t+\Delta t$ from the distribution $W(\boldsymbol{R}, t)$ at the earlier time t. In view of (108) and its independence of $\boldsymbol{R}$ we have the integral equation

$$W(\boldsymbol{R}, t+\Delta t)=\int_{-\infty}^{+\infty}W(\boldsymbol{R}-\Delta\boldsymbol{R}, t)\psi(\Delta\boldsymbol{R};\Delta t)d(\Delta\boldsymbol{R}). \tag{109}$$

Since $\langle|\Delta\boldsymbol{R}|^2\rangle_{\text{Av}}$ is assumed to be small we can expand $W(\boldsymbol{R}-\Delta\boldsymbol{R}, t)$ under the integral sign in (109) in a Taylor series and integrate term by term. We find

$$\left.\begin{aligned}W(\boldsymbol{R}, t+\Delta t)=&\frac{1}{(4\pi D\Delta t)^{\frac{3}{2}}}\int_{-\infty}^{+\infty}\int_{-\infty}^{+\infty}\int_{-\infty}^{+\infty}\exp\left(-|\Delta\boldsymbol{R}|^2/4D\Delta t\right)\Bigg\{W(\boldsymbol{R}, t)-\Delta X\frac{\partial W}{\partial X}-\Delta Y\frac{\partial W}{\partial Y}\\&-\Delta Z\frac{\partial W}{\partial Z}+\frac{1}{2}\Bigg[(\Delta X)^2\frac{\partial^2 W}{\partial X^2}+(\Delta Y)^2\frac{\partial^2 W}{\partial Y^2}+(\Delta Z)^2\frac{\partial^2 W}{\partial Z^2}+2\Delta X\Delta Y\frac{\partial^2 W}{\partial X\partial Y}\\&+2\Delta Y\Delta Z\frac{\partial^2 W}{\partial Y\partial Z}+2\Delta Z\Delta X\frac{\partial^2 W}{\partial Z\partial X}\Bigg]+\cdots\Bigg\}d(\Delta X)d(\Delta Y)d(\Delta Z)\\=&\,W(\boldsymbol{R}, t)+D\Delta t\left(\frac{\partial^2 W}{\partial X^2}+\frac{\partial^2 W}{\partial Y^2}+\frac{\partial^2 W}{\partial Z^2}\right)+O([\Delta t]^2).\end{aligned}\right\} \tag{110}$$

$$= e^{D\Delta t\,\nabla_R^2}\,W(\underline{R}, t)$$

Accordingly,

$$\frac{\partial W}{\partial t}\Delta t+O([\Delta t]^2)=D\left(\frac{\partial^2 W}{\partial X^2}+\frac{\partial^2 W}{\partial Y^2}+\frac{\partial^2 W}{\partial Z^2}\right)\Delta t+O([\Delta t]^2). \tag{111}$$

Passing now to the limit of $\Delta t=0$ we obtain

$$\frac{\partial W}{\partial t}=D\left(\frac{\partial^2 W}{\partial X^2}+\frac{\partial^2 W}{\partial Y^2}+\frac{\partial^2 W}{\partial Z^2}\right) \tag{112}$$

which is the required differential equation. And, it is seen that $W(\boldsymbol{R}, t)$ defined according to Eq. (107) is indeed the fundamental solution of this differential equation.

Equation (112) is the standard form of the *equation of diffusion* or of heat conduction. This analogy that exists between our differential Eq. (112) to the equation of diffusion provides a new interpretation of the problem of random flights in terms of a *diffusion coefficient D*.

It is well known that in the *macroscopic* theory of diffusion if $W(\boldsymbol{R}, t)$ denotes the concentration of the diffusing substance at $\boldsymbol{R}$ and at time t, then the amount crossing an area $\Delta\sigma$ in time Δt is given by

$$-D(\mathbf{1}_{\Delta\sigma}\cdot\text{grad } W)\Delta\sigma\Delta t, \tag{113}$$

where $\mathbf{1}_{\Delta\sigma}$ is a unit vector normal to the element of area $\Delta\sigma$. The diffusion equation is an elementary consequence of this fact. Consequently, we may describe the motion of a large number of particles describing random flights without mutual interference as a process of diffusion with the diffusion coefficient

$$D=n\langle r^2\rangle_{\text{Av}}/6. \tag{114}$$

With this visualization of the problem, the boundary conditions

$$W=0 \text{ on an element of surface which is a perfect absorber} \tag{115}$$

and

$$\text{grad } W=0 \text{ normal to an element surface which is a perfect reflector} \tag{116}$$

become intelligible. Further, according to Eq. (113), the rate at which particles appear on an absorbing screen per unit area, and per unit time, is given by

$$-D(\mathbf{1}\cdot\text{grad } W)_{W=0} \tag{117}$$

where $\mathbf{1}$ is a unit vector normal to the absorbing surface. This is in agreement with Eq. (33).

We shall now derive the differential equation for the problem of random flights in its general form considered in §4, subsection (d). This problem differs from the one we have just considered in that the probability distribution $\tau(\boldsymbol{r})$ governing the individual displacements $\boldsymbol{r}$ is now a function with no special symmetry properties. Accordingly, the first moments of τ cannot be assumed to vanish; further, the second moments define a general symmetric tensor of the second rank. Under these circumstances, the probability of finding the particle between $\boldsymbol{R}$ and $\boldsymbol{R}+d\boldsymbol{R}$ after it has suffered a large number of displacements is given by [cf. Eq. (103)]

$$W(\boldsymbol{R})d\boldsymbol{R}=\frac{1}{(8\pi^3N^3\langle x^2\rangle\langle y^2\rangle\langle z^2\rangle)^{\frac{1}{2}}}\exp\left[-\frac{(X-N\langle x\rangle)^2}{2N\langle x^2\rangle}-\frac{(Y-N\langle y\rangle)^2}{2N\langle y^2\rangle}-\frac{(Z-N\langle z\rangle)^2}{2N\langle z^2\rangle}\right]d\boldsymbol{R}. \tag{118}$$

In writing the probability distribution $W(\boldsymbol{R})$ in this form we have supposed that the coordinate system has been so chosen that the X, Y, and Z directions are along the principal axes of the moment ellipsoid.

Assuming that, on the average, the particle suffers n displacements per unit time we can rewrite our expression for $W(\boldsymbol{R})$ more conveniently in the form

$$W(\boldsymbol{R})=\frac{1}{8(\pi t)^{\frac{3}{2}}(D_1D_2D_3)^{\frac{1}{2}}}\exp\left[-\frac{(X+\beta_1 t)^2}{4D_1t}-\frac{(Y+\beta_2 t)^2}{4D_2t}-\frac{(Z+\beta_3 t)^2}{4D_3t}\right] \tag{119}$$

where we have written

$$\left.\begin{aligned}&D_1=\tfrac{1}{2}n\langle x^2\rangle;\quad D_2=\tfrac{1}{2}n\langle y^2\rangle;\quad D_3=\tfrac{1}{2}n\langle z^2\rangle,\\&\beta_1=-n\langle x\rangle;\quad \beta_2=-n\langle y\rangle;\quad \beta_3=-n\langle z\rangle.\end{aligned}\right\} \tag{120}$$

To make the passage to a differential equation, we consider, as before, an interval Δt which is long enough for the particle to suffer a large number of individual displacements but short enough for the mean square increment $\langle|\Delta\boldsymbol{R}|^2\rangle_{\text{Av}}$ to be small. The probability that the particle suffers an increment $\Delta\boldsymbol{R}$ in the interval Δt is therefore governed by the distribution function

$$\psi(\Delta\boldsymbol{R};\Delta t)=\frac{1}{8(\pi\Delta t)^{\frac{3}{2}}(D_1D_2D_3)^{\frac{1}{2}}}\exp\left[-\frac{(\Delta X+\beta_1\Delta t)^2}{4D_1\Delta t}-\frac{(\Delta Y+\beta_2\Delta t)^2}{4D_2\Delta t}-\frac{(\Delta Z+\beta_3\Delta t)^2}{4D_3\Delta t}\right]. \tag{121}$$

Hence, analogous to Eqs. (109) and (110) we now have

$$\left.\begin{aligned}W(\boldsymbol{R},t+\Delta t)&=W(\boldsymbol{R},t)+\frac{\partial W}{\partial t}\Delta t+O([\Delta t]^2)=\int_{-\infty}^{+\infty}W(\boldsymbol{R}-\Delta\boldsymbol{R},t)\psi(\Delta\boldsymbol{R};\Delta t)d(\Delta\boldsymbol{R})\\&=\frac{1}{8(\pi\Delta t)^{\frac{3}{2}}(D_1D_2D_3)^{\frac{1}{2}}}\int_{-\infty}^{+\infty}\int_{-\infty}^{+\infty}\int_{-\infty}^{+\infty}\exp\left[-\frac{(\Delta X+\beta_1\Delta t)^2}{4D_1\Delta t}-\frac{(\Delta Y+\beta_2\Delta t)^2}{4D_2\Delta t}\right.\\&\quad\left.-\frac{(\Delta Z+\beta_3\Delta t)^2}{4D_3\Delta t}\right]\left\{W(\boldsymbol{R},t)-\left(\Delta X\frac{\partial W}{\partial X}+\Delta Y\frac{\partial W}{\partial Y}+\Delta Z\frac{\partial W}{\partial Z}\right)\right.\\&\quad+\frac{1}{2}\left(\Delta X^2\frac{\partial^2W}{\partial X^2}+\Delta Y^2\frac{\partial^2W}{\partial Y^2}+\Delta Z^2\frac{\partial^2W}{\partial Z^2}+2\Delta X\Delta Y\frac{\partial^2W}{\partial X\partial Y}+2\Delta Y\Delta Z\frac{\partial^2W}{\partial Y\partial Z}\right.\\&\quad\left.\left.+2\Delta Z\Delta X\frac{\partial^2W}{\partial Z\partial X}\right)-\cdots\right\}d(\Delta X)d(\Delta Y)d(\Delta Z).\end{aligned}\right\} \tag{122}$$

Since for the distribution function (121)

$$\langle\Delta X\rangle_{\text{Av}}=-\beta_1\Delta t;\quad\langle\Delta Y\rangle_{\text{Av}}=-\beta_2\Delta t;\quad\langle\Delta Z\rangle_{\text{Av}}=-\beta_3\Delta t, \tag{123}$$

and

$$\left.\begin{aligned}&\langle\Delta X^2\rangle_{\text{Av}}=2D_1\Delta t+\beta_1^2\Delta t^2;\quad\langle\Delta Y\Delta Z\rangle_{\text{Av}}=\beta_2\beta_3\Delta t^2,\\&\langle\Delta Y^2\rangle_{\text{Av}}=2D_2\Delta t+\beta_2^2\Delta t^2;\quad\langle\Delta Z\Delta X\rangle_{\text{Av}}=\beta_3\beta_1\Delta t^2,\\&\langle\Delta Z^2\rangle_{\text{Av}}=2D_3\Delta t+\beta_3^2\Delta t^2;\quad\langle\Delta X\Delta Y\rangle_{\text{Av}}=\beta_1\beta_2\Delta t^2,\end{aligned}\right\} \tag{124}$$

we conclude from Eq. (122) that

$$\frac{\partial W}{\partial t}\Delta t+O([\Delta t]^2)=\left(\beta_1\frac{\partial W}{\partial X}+\beta_2\frac{\partial W}{\partial Y}+\beta_3\frac{\partial W}{\partial Z}\right)\Delta t+\left(D_1\frac{\partial^2W}{\partial X^2}+D_2\frac{\partial^2W}{\partial Y^2}+D_3\frac{\partial^2W}{\partial Z^2}\right)\Delta t+O([\Delta t]^2). \tag{125}$$

Passing now to the limit $\Delta t=0$ we obtain

$$\frac{\partial W}{\partial t}=\beta_1\frac{\partial W}{\partial X}+\beta_2\frac{\partial W}{\partial Y}+\beta_3\frac{\partial W}{\partial Z}+D_1\frac{\partial^2W}{\partial X^2}+D_2\frac{\partial^2W}{\partial Y^2}+D_3\frac{\partial^2W}{\partial Z^2}, \tag{126}$$

which is the required differential equation. According to this equation we can describe the phenomenon under discussion as a general process of diffusion in which the number of particles crossing elements of area normal to the X, Y, and Z direction per unit area and per unit time are given, respectively, by

$$-\beta_1 W - D_1\frac{\partial W}{\partial X};\quad -\beta_2 W - D_2\frac{\partial W}{\partial Y};\quad -\beta_3 W - D_3\frac{\partial W}{\partial Z}. \tag{127}$$

For the purposes of solving the differential Eq. (126) it is convenient to introduce a change in the independent variable. Let

$$W = U \exp\left[-\frac{\beta_1}{2D_1}(X-X_0)-\frac{\beta_2}{2D_2}(Y-Y_0)-\frac{\beta_3}{2D_3}(Z-Z_0)-\frac{\beta_1^2}{4D_1}t-\frac{\beta_2^2}{4D_2}t-\frac{\beta_3^2}{4D_3}t\right]. \tag{128}$$

We verify that Eq. (126) now reduces to

$$\frac{\partial U}{\partial t} = D_1\frac{\partial^2 U}{\partial X^2}+D_2\frac{\partial^2 U}{\partial Y^2}+D_3\frac{\partial^2 U}{\partial Z^2}. \tag{129}$$

The fundamental solution of this differential equation is

$$U = \frac{\text{Constant}}{(D_1D_2D_3t^3)^{\frac{1}{2}}}\exp\left[-\frac{(X-X_0)^2}{4D_1t}-\frac{(Y-Y_0)^2}{4D_2t}-\frac{(Z-Z_0)^2}{4D_3t}\right]. \tag{130}$$

Returning to the variable W, we have

$$W = \frac{\text{Constant}}{(D_1D_2D_3t^3)^{\frac{1}{2}}}\exp\left[-\frac{(X-X_0+\beta_1t)^2}{4D_1t}-\frac{(Y-Y_0+\beta_2t)^2}{4D_2t}-\frac{(Z-Z_0+\beta_3t)^2}{4D_3t}\right]. \tag{131}$$

In other words, the distribution (119) does indeed represent the fundamental solution of the differential Eq. (126).

CHAPTER II

THE THEORY OF THE BROWNIAN MOTION

1. Introductory Remarks. Langevin's Equation

In the studies on Brownian motion we are principally concerned with the perpetual irregular motions exhibited by small grains or particles of colloidal size immersed in a fluid. As is now well known, we witness in Brownian movement the phenomenon of molecular agitation on a reduced scale by particles very large on the molecular scale—so large in fact as to be readily visible in an ultramicroscope. The perpetual motions of the Brownian particles are maintained by fluctuations in the collisions with the molecules of the surrounding fluid. Under normal conditions, in a liquid, a Brownian particle will suffer about 10^{21} collisions per second and this is so frequent that we cannot really speak of separate collisions. Also, since each collision can be thought of as producing a kink in the path of the particle, it follows that we cannot hope to follow the path in any detail—indeed, to our senses the details of the path are impossibly fine.

The modern theory of the Brownian motion of a *free particle* (i.e., in the absence of an external field of force) generally starts with Langevin's equation

$$d\mathbf{u}/dt = -\beta\mathbf{u} + \mathbf{A}(t), \tag{132}$$

where $\mathbf{u}$ denotes the velocity of the particle. According to this equation, the influence of the surrounding medium on the motion of the particle can be split up into two parts: first, a systematic part $-\beta\mathbf{u}$ representing a *dynamical friction* experienced by the particle and second, a fluctuating part $\mathbf{A}(t)$ which is characteristic of the Brownian motion.

Regarding the frictional term $-\beta \boldsymbol{u}$ it is assumed that this is governed by Stokes' law which states that the frictional force decelerating a spherical particle of radius a and mass m is given by $6\pi a\eta \boldsymbol{u}/m$ where η denotes the coefficient of viscosity of the surrounding fluid. Hence

$$\beta = 6\pi a\eta/m. \tag{133}$$

As for the fluctuating part $\boldsymbol{A}(t)$ the following principal assumptions are made:

(i) $\boldsymbol{A}(t)$ is independent of $\boldsymbol{u}$.
(ii) $\boldsymbol{A}(t)$ varies extremely rapidly compared to the variations of $\boldsymbol{u}$.

The second assumption implies that time intervals of duration Δt exist such that during Δt the variations in $\boldsymbol{u}$ that are to be expected are very small indeed while during the same interval $\boldsymbol{A}(t)$ may undergo several fluctuations. Alternatively, we may say that though $\boldsymbol{u}(t)$ and $\boldsymbol{u}(t+\Delta t)$ are expected to differ by a negligible amount, no correlation between $\boldsymbol{A}(t)$ and $\boldsymbol{A}(t+\Delta t)$ exists. (The assumptions which are made here are quite analogous to those made in Chapter I, §5 in the passage to the differential equation for the problem of random flights; also see §§2 and 4 in this chapter.)

We shall show in the following sections how with the assumptions made in the foregoing paragraphs, we can derive from Langevin's equation all the physically significant relations concerning the motions of the Brownian particles. But we should draw attention even at this stage to the very drastic nature of assumptions implicit in the very writing of an equation of the form (132). For we have in reality supposed that we can divide the phenomenon into two parts, one in which the discontinuity of the events taking place is essential while in the other it is trivial and can be ignored. In view of the discontinuities in all matter and all events, this is a prima facie, an *ad-hoc* assumption. They are however made with reliance on physical intuition and the *aposteriori* justification by the success of the hypothesis. However, the correct procedure would be to treat the phenomenon in its entirety without appealing to the laws of continuous physics except insofar as they can be explicitly justified. As we shall see in Chapter IV a problem which occurs in stellar dynamics appears to provide a model in which the rigorous procedure can be explicitly followed.

2. The Theory of the Brownian Motion of a Free Particle

Our problem is to solve the stochastic differential equation (132) subject to the restrictions on $\boldsymbol{A}(t)$ stated in the preceding section. But "solving" a stochastic differential equation like (132) is not the same thing as solving any ordinary differential equation. For one thing, Eq. (132) involves the function $\boldsymbol{A}(t)$ which, as we shall presently see, has only statistically defined properties. Consequently, "solving" the Langevin Eq. (132) has to be understood rather in the sense of specifying a probability distribution $W(\boldsymbol{u}, t; \boldsymbol{u}_0)$ which governs the probability of occurrence of the velocity $\boldsymbol{u}$ at time t given that $\boldsymbol{u}=\boldsymbol{u}_0$ at $t=0$. Of this function $W(\boldsymbol{u}, t; \boldsymbol{u}_0)$ we should clearly require that, as $t\to 0$,

$$W(\boldsymbol{u}, t; \boldsymbol{u}_0)\to\delta(u_x-u_{x,0})\delta(u_y-u_{y,0})\delta(u_z-u_{z,0}) \quad (t\to 0), \tag{134}$$

where the δ's are Dirac's δ functions. Further, the physical circumstances of the problem require that we demand of $W(\boldsymbol{u}, t; \boldsymbol{u}_0)$ that it tend to a Maxwellian distribution for the temperature T of the surrounding fluid, *independently* of $\boldsymbol{u}_0$ as $t\to\infty$:

$$W(\boldsymbol{u}, t; \boldsymbol{u}_0)\to\left(\frac{m}{2\pi kT}\right)^{\frac{3}{2}} \exp\left(-m|\boldsymbol{u}|^2/2kT\right) \quad (t\to\infty). \tag{135}$$

This last demand on $W(\boldsymbol{u}, t; \boldsymbol{u}_0)$ conversely requires that $\boldsymbol{A}(t)$ satisfy certain statistical requirements. For, according to the Langevin equation we have the formal solution

$$\boldsymbol{u}-\boldsymbol{u}_0 e^{-\beta t}=e^{-\beta t}\int_0^t e^{\beta\xi}\boldsymbol{A}(\xi)d\xi. \tag{136}$$

Consequently, the statistical properties of

$$u - u_0 e^{-\beta t} \tag{137}$$

must be the same as those of

$$e^{-\beta t}\int_0^t e^{\beta\xi} A(\xi) d\xi. \tag{138}$$

And, as $t \to \infty$ the quantity (137) tends to u; hence the distribution of

$$\underset{t\to\infty}{\text{Limit}} \left\{ e^{-\beta t}\int_0^t e^{\beta\xi} A(\xi) d\xi \right\} \tag{139}$$

must be the Maxwellian distribution

$$(m/2\pi kT)^{\frac{3}{2}} \exp(-m|u|^2/2kT). \tag{140}$$

Now one of our principal assumptions concerning $A(t)$ is that it varies extremely rapidly compared to any of the other quantities which enter into our discussion. Further, the fluctuating acceleration experienced by the Brownian particles is statistical in character in the sense that Brownian particles having the same initial coordinates and/or velocities will suffer accelerations which will differ from particle to particle both in magnitude and in their dependence on time. However, on account of the rapidity of these fluctuations, we can always divide an interval of time which is long enough for any of the physical parameters like the position or the velocity of a Brownian particle to change appreciably, into a very large number of subintervals of duration Δt such that during each of these subintervals we can treat all functions of time except $A(t)$ which enter in our formulae as constants. Thus, the quantity on the right-hand side of Eq. (136) can be written as

$$e^{-\beta t}\sum_j e^{\beta j\Delta t}\int_{j\Delta t}^{(j+1)\Delta t} A(\xi) d\xi. \tag{141}$$

Let

$$B(\Delta t) = \int_t^{t+\Delta t} A(\xi) d\xi. \tag{142}$$

The physical meaning of $B(\Delta t)$ is that it represents the net acceleration which a Brownian particle may suffer on a given occasion during an interval of time Δt.

Equation (136) becomes

$$u - u_0 e^{-\beta t} = \sum_j e^{\beta(j\Delta t - t)} B(\Delta t), \tag{143}$$

and we require that as $t \to \infty$ the quantity on the right-hand side tends to the Maxwellian distribution (140). We now assert that this requires *the probability of occurrence of different values for* $B(\Delta t)$ *be governed by the distribution function*

$$w(B[\Delta t]) = \frac{1}{(4\pi q\Delta t)^{\frac{3}{2}}} \exp(-|B(\Delta t)|^2/4q\Delta t) \tag{144}$$

where

$$q = \beta kT/m. \tag{145}$$

To prove this assertion we have to show that the distribution function $W(u, t; u_0)$ derived on the basis of Eqs. (143) and (144) does in fact tend to the Maxwellian distribution (140) as $t \to \infty$. We shall presently show that this is the case but we may remark meantime on the formal similarity of Eq. (144) giving the probability distribution of the acceleration $B(\Delta t)$ suffered by a Brownian particle in time Δt and Eq. (108) giving the probability distribution of the increment ΔR in the position of a particle describing random flights in time Δt. It will be recalled that for the validity of Eq. (108) it is necessary that Δt be long enough for a large number of individual displacements to occur; analogously, our expression for $w(\boldsymbol{B}[\Delta t])$ is valid only for times Δt large compared to the average period of a single fluctuation of $\boldsymbol{A}(t)$. Now, the period of fluctuation of $\boldsymbol{A}(t)$ is clearly of the order of the time between successive collisions between the Brownian particle and the molecules of the surrounding fluid; in a liquid this is generally of the order of 10^{-21} sec. Accordingly, the similarity of our expression for $w(\boldsymbol{B}[\Delta t])$ with Eq. (108) in the theory of random flights, leads us to interpret the acceleration $\boldsymbol{B}(\Delta t)$ suffered by a Brownian particle (in a time Δt large compared with the frequency of collisions with the surrounding molecules) as the result of superposition of the large number of random accelerations caused by collisions with the individual molecules. This is of course eminently reasonable; but the reason why q in Eq. (144) has to be precisely that given by Eq. (145) is due to our requirement that $W(\boldsymbol{u}, t; \boldsymbol{u}_0)$ tend to the Maxwellian distribution (140) as $t \to \infty$. We shall return to these questions again in §5.

We now proceed to prove our assertion concerning Eqs. (143), (144) and (145):

We first prove the following lemma:

Lemma I. Let

$$\boldsymbol{R} = \int_0^t \psi(\xi)\boldsymbol{A}(\xi)d\xi. \tag{146}$$

Then, the probability distribution of $\boldsymbol{R}$ *is given by*

$$W(\boldsymbol{R}) = \frac{1}{\left[4\pi q \int_0^t \psi^2(\xi)d\xi\right]^{\frac{3}{2}}} \exp\left(-|\boldsymbol{R}|^2 \Big/ 4q\int_0^t \psi^2(\xi)d\xi\right). \tag{147}$$

In order to prove this, we first divide the interval $(0, t)$ into a large number of subintervals of duration Δt. We can then write

$$\boldsymbol{R} = \sum_j \psi(j\Delta t) \int_{j\Delta t}^{(j+1)\Delta t} \boldsymbol{A}(\xi)d\xi. \tag{148}$$

Remembering our definition of $\boldsymbol{B}(\Delta t)$ [Eq. (142)] we can express $\boldsymbol{R}$ in the form

$$\boldsymbol{R} = \sum\nolimits_j \boldsymbol{r}_j, \tag{149}$$

where

$$\boldsymbol{r}_j = \psi(j\Delta t)\boldsymbol{B}(\Delta t) = \psi_j \boldsymbol{B}(\Delta t). \tag{150}$$

According to Eq. (144) the probability distribution of $\boldsymbol{r}_j$ is governed by

$$\tau(\boldsymbol{r}_j) = \frac{1}{(2\pi l_j^2/3)^{\frac{3}{2}}} \exp(-3|\boldsymbol{r}_j|^2/2l_j^2), \tag{151}$$

where we have written

$$l_j^2 = 6q\psi_j^2\Delta t. \tag{152}$$

A comparison of Eqs. (149) and (151) with Eqs. (34) and (57) shows that we have reduced our present problem to the special case in the theory of random flights considered in Chapter **I**, §4 case (a). Hence, [cf. Eqs. (59) and (62)]

$$W(\boldsymbol{R}) = \frac{1}{(2\pi \sum l_j^2/3)^{\frac{3}{2}}} \exp(-3|\boldsymbol{R}|^2/2\sum l_j^2). \tag{153}$$

But

$$\left.\begin{aligned} \sum l_j^2 &= 6q \sum\nolimits_j \psi_j^2 \Delta t = 6q \sum\nolimits_j \psi^2(j\Delta t)\Delta t, \\ &= 6q\int_0^t \psi^2(\xi)d\xi. \end{aligned}\right\} \tag{154}$$

We therefore have

$$W(\mathbf{R})=\frac{1}{\left[4\pi q\int_0^t \psi^2(\xi)d\xi\right]^{\frac{3}{2}}}\exp\left(-|\mathbf{R}|^2\Big/4q\int_0^t\psi^2(\xi)d\xi\right), \tag{155}$$

which proves the lemma.

Returning to Eq. (136) we notice that we can express the right-hand side of this equation in the form

$$\int_0^t \psi(\xi)\mathbf{A}(\xi)d\xi \tag{156}$$

if we define

$$\psi(\xi)=e^{\beta(\xi-t)}. \tag{157}$$

We can therefore apply lemma I and with the foregoing definition of $\psi(\xi)$, Eq. (155) governs the probability distribution of

$$\mathbf{u}-\mathbf{u}_0e^{-\beta t}. \tag{158}$$

Since, now,

$$\int_0^t \psi^2(\xi)d\xi=\int_0^t e^{2\beta(\xi-t)}d\xi=\frac{1}{2\beta}(1-e^{-2\beta t}), \tag{159}$$

and remembering that according to Eq. (145)

$$q/\beta=kT/m \tag{160}$$

we have proved that

$$W(\mathbf{u},t;\mathbf{u}_0)=\left[\frac{m}{2\pi kT(1-e^{-2\beta t})}\right]^{\frac{3}{2}}\exp\left[-m|\mathbf{u}-\mathbf{u}_0e^{-\beta t}|^2/2kT(1-e^{-2\beta t})\right]. \tag{161}$$

We verify that according to this equation

$$W(\mathbf{u},t;\mathbf{u}_0)\rightarrow\left(\frac{m}{2\pi kT}\right)^{\frac{3}{2}}\exp\left(-m|\mathbf{u}|^2/2kT\right)\quad(t\rightarrow\infty) \tag{162}$$

i.e., the Maxwellian distribution (140). This proves the assertion we made that with the statistical properties of $\mathbf{B}(\Delta t)$ implied in Eqs. (144) and (145), Eq. (143) leads to a distribution $W(\mathbf{u},t;\mathbf{u}_0)$ which tends to be Maxwellian independent of $\mathbf{u}_0$ as $t\rightarrow\infty$.

We shall now show how with the assumptions already made concerning $\mathbf{B}(\Delta t)$ we can further derive the distribution of the displacement $\mathbf{r}$ of a Brownian particle at time t given that the particle is at $\mathbf{r}_0$ with a velocity $\mathbf{u}_0$ at time $t=0$:

Since

$$\mathbf{r}-\mathbf{r}_0=\int_0^t \mathbf{u}(t)dt, \tag{163}$$

we have according to Eq. (136)

$$\mathbf{r}-\mathbf{r}_0=\int_0^t d\eta\left\{\mathbf{u}_0e^{-\beta\eta}+e^{-\beta\eta}\int_0^\eta d\xi e^{\beta\xi}\mathbf{A}(\xi)\right\} \tag{164}$$

or

$$\mathbf{r}-\mathbf{r}_0-\beta^{-1}\mathbf{u}_0(1-e^{-\beta t})=\int_0^t d\eta e^{-\beta\eta}\int_0^\eta d\xi e^{\beta\xi}\mathbf{A}(\xi). \tag{165}$$

We can simplify the right-hand side of this equation by an integration by parts. We find

$$\mathbf{r}-\mathbf{r}_0-\beta^{-1}\mathbf{u}_0(1-e^{-\beta t})=-\beta^{-1}e^{-\beta t}\int_0^t e^{\beta\xi}\mathbf{A}(\xi)d\xi+\beta^{-1}\int_0^t \mathbf{A}(\xi)d\xi. \tag{166}$$

Again, we can reduce this equation to the form

$$\mathbf{r}-\mathbf{r}_0-\beta^{-1}\mathbf{u}_0(1-e^{-\beta t})=\int_0^t \psi(\xi)\mathbf{A}(\xi)d\xi, \tag{167}$$

by defining

$$\psi(\xi)=\beta^{-1}(1-e^{\beta(\xi-t)}). \tag{168}$$

Thus lemma I can be applied and with the definition of $\psi(\xi)$ according to Eq. (168), Eq. (155) governs the probability distribution of

$$\mathbf{r}-\mathbf{r}_0-\beta^{-1}\mathbf{u}_0(1-e^{-\beta t}) \tag{169}$$

i.e., of $\mathbf{r}$ at time t for given $\mathbf{r}_0$ and $\mathbf{u}_0$. Since,

$$\left.\begin{aligned}\int_0^t \psi^2(\xi)d\xi&=\frac{1}{\beta^2}\int_0^t(1-e^{\beta(\xi-t)})^2d\xi,\\ &=\frac{1}{2\beta^3}(2\beta t-3+4e^{-\beta t}-e^{-2\beta t}),\end{aligned}\right\} \tag{170}$$

we have

$$W(\mathbf{r},t;\mathbf{r}_0,\mathbf{u}_0)=\left\{\frac{m\beta^2}{2\pi kT[2\beta t-3+4e^{-\beta t}-e^{-2\beta t}]}\right\}^{\frac{3}{2}}\exp -\left\{\frac{m\beta^2|\mathbf{r}-\mathbf{r}_0-\mathbf{u}_0(1-e^{-\beta t})/\beta|^2}{2kT[2\beta t-3+4e^{-\beta t}-e^{-2\beta t}]}\right\}. \tag{171}$$

For intervals of time long compared to β^{-1} the foregoing expression simplifies considerably. For, under these circumstances we can ignore the exponential and the constant terms as compared to $2\beta t$. Further, as we shall presently show, $\langle|\mathbf{r}-\mathbf{r}_0|^2\rangle_{\text{Av}}$ is of order t [cf. Eq. (174)]; hence we can also neglect $\mathbf{u}_0(1-e^{-\beta t})\beta^{-1}$ compared to $\mathbf{r}-\mathbf{r}_0$. Thus Eq. (171) reduces to

$$W(\mathbf{r},t;\mathbf{r}_0,\mathbf{u}_0)\simeq\frac{1}{(4\pi Dt)^{\frac{3}{2}}}\exp\left(-|\mathbf{r}-\mathbf{r}_0|^2/4Dt\right)\quad(t\gg\beta^{-1}) \tag{172}$$

where we have introduced the "diffusion coefficient" D defined by

$$D=kT/m\beta=kT/6\pi a\eta. \tag{173}$$

In Eq. (173) we have substituted for β according to Eq. (133).

From Eq. (172) we obtain for the mean square displacement along any given direction (say, the x direction) the formula

$$\langle(x-x_0)^2\rangle_{\text{Av}}=\tfrac{1}{3}\langle|\mathbf{r}-\mathbf{r}_0|^2\rangle_{\text{Av}}=2Dt=(kT/3\pi a\eta)t. \tag{174}$$

This is Einstein's result. Equation (174) has been verified by Perrin to lead to consistent and satisfactory values for the Boltzmann constant k by observation of $\langle(x-x_0^2)\rangle_{\text{Av}}/t$ over wide ranges of T, η and a.

The law of distribution of displacements (172) has been exhaustively tested by observation. Perrin gives the following sets of counts of the displacements of a grain of radius 2.1×10^{-5} cm at 30 sec. intervals. Out of a number N of such observations the number of observed values of x displacements between x_1 and x_2 should be

$$\frac{N}{\pi^{\frac{1}{2}}}\int_{x_1}^{x_2}\exp\left(-x^2/4Dt\right)\frac{dx}{(4Dt)^{\frac{1}{2}}}.$$

The agreement is satisfactory. See Table II.

Comparing Eq. (172) with the solution for the problem of random flights obtained in Eq. (107) we conclude that for times $t\gg\beta^{-1}$ we can regard the motion of a Brownian particle as one of random

TABLE II. Observations and calculations of the distribution of the displacements of a Brownian particle.

Range $x\times10^4$ cm	1st set Obs.	1st set Calc.	2nd set Obs.	2nd set Calc.	Total Obs.	Total Calc.
0 - 3.4	82	91	86	84	168	175
3.4- 6.8	66	70	65	63	131	132
6.8-10.2	46	39	31	36	77	75
10.2-17.0	27	23	23	21	50	44

flights. And therefore, according to the ideas of *I* §5, describe the motion of Brownian particles also as one of diffusion and governed by the diffusion equation. We shall return to this connection with the diffusion equation from a more general point of view in §4.

Returning to Eq. (171) we see that, quite generally, we have

$$\langle|\mathbf{r}-\mathbf{r}_0|^2\rangle_{\text{Av}}=\frac{|\mathbf{u}_0|^2}{\beta^2}(1-e^{-\beta t})^2+3\frac{kT}{m\beta^2}(2\beta t-3+4e^{-\beta t}-e^{-2\beta t}). \tag{175}$$

Averaging this equation over all values of $\mathbf{u}_0$ and remembering that $\langle|\mathbf{u}_0|^2\rangle_{\text{Av}}=3kT/m$ we obtain

$$\langle\langle|\mathbf{r}-\mathbf{r}_0|^2\rangle\rangle_{\text{Av}}=6\frac{kT}{m\beta^2}(\beta t-1+e^{-\beta t}). \tag{175'}$$

For $t\rightarrow\infty$, Eq. (175′) is in agreement with our result (174), while for $t\rightarrow0$ we have instead

$$\langle\langle|\mathbf{r}-\mathbf{r}_0|^2\rangle\rangle_{\text{Av}}=3\frac{kT}{m}t^2=\langle|\mathbf{u}_0|^2\rangle_{\text{Av}}t^2. \tag{175''}$$

So far we have only inquired into the law of distributions of $\mathbf{u}$ and $\mathbf{r}$ separately. But we can also ask for the distribution $W(\mathbf{r}, \mathbf{u}, t; \mathbf{u}_0, \mathbf{r}_0)$ governing the probability of the simultaneous occurrence of the velocity $\mathbf{u}$ and the position $\mathbf{r}$ at time t, given that $\mathbf{u}=\mathbf{u}_0$ and $\mathbf{r}=\mathbf{r}_0$ at $t=0$. The solution to this problem can be obtained from the following lemma:

Lemma II. *Let*

$$\mathbf{R}=\int_0^t\psi(\xi)\mathbf{A}(\xi)d\xi, \tag{176}$$

and

$$\mathbf{S}=\int_0^t\phi(\xi)\mathbf{A}(\xi)d\xi. \tag{177}$$

Then, the bivariate probability distribution of **R** *and* **S** *is given by*

$$W(\mathbf{R}, \mathbf{S})=\frac{1}{8\pi^3(FG-H^2)^{\frac{3}{2}}}\exp\left[-(G|\mathbf{R}|^2-2H\mathbf{R}\cdot\mathbf{S}+F|\mathbf{S}|^2)/2(FG-H^2)\right] \tag{178}$$

where

$$F=2q\int_0^t\psi^2(\xi)d\xi;\quad G=2q\int_0^t\phi^2(\xi)d\xi;\quad H=2q\int_0^t\phi(\xi)\psi(\xi)d\xi. \tag{179}$$

The lemma is proved by writing **R** and **S** in the forms [cf. Eqs. (149) and (150)]

$$\mathbf{R}=\sum_j\psi(j\Delta t)\mathbf{B}(\Delta t);\quad \mathbf{S}=\sum_j\phi(j\Delta t)\mathbf{B}(\Delta t) \tag{180}$$

and remembering that the distribution of **B** is Gaussian according to Eq. (144). The problem then reduces to the one considered in Appendix II and the solution stated readily follows.

To obtain the distribution $W(\mathbf{r}, \mathbf{u}, t; \mathbf{u}_0, \mathbf{r}_0)$ we have only to set [cf. Eqs. (157), (158), (167) and (168)]

$$\left.\begin{aligned} \mathbf{R}&=\mathbf{r}-\mathbf{r}_0-\beta^{-1}\mathbf{u}_0(1-e^{-\beta t}); \quad \psi(\xi)=\beta^{-1}(1-e^{\beta(\xi-t)}),\\ \mathbf{S}&=\mathbf{u}-\mathbf{u}_0e^{-\beta t}; \quad \phi(\xi)=e^{\beta(\xi-t)}, \end{aligned}\right\} \tag{181}$$

and [cf. Eqs. (159) and (170)]

$$F=q\beta^{-3}(2\beta t-3+4e^{-\beta t}-e^{-2\beta t}); \quad G=q\beta^{-1}(1-e^{-2\beta t}), \tag{182}$$

and finally

$$H=2q\beta^{-1}\int_0^t e^{\beta(\xi-t)}(1-e^{\beta(\xi-t)})dt=q\beta^{-2}(1-e^{-\beta t})^2. \tag{183}$$

3. The Theory of the Brownian Motion of a Particle in a Field of Force. The Harmonically Bound Particle

In the presence of an external field of force, the Langevin Eq. (132) is generalized to

$$d\mathbf{u}/dt=-\beta\mathbf{u}+\mathbf{A}(t)+\mathbf{K}(\mathbf{r},t) \tag{184}$$

where $\mathbf{K}(\mathbf{r}, t)$ is the acceleration produced by the field. In writing this equation we are making the same general assumptions as are involved in writing the original Langevin equation (cf. the remarks at the end of §1).

In solving the stochastic equation (184) we attribute to $\mathbf{A}(t)$ or more particularly for

$$\mathbf{B}(\Delta t)=\int_t^{t+\Delta t}\mathbf{A}(\xi)d\xi \tag{185}$$

the statistical properties already assigned in the preceding section [Eq. (144)]. The method of solution is illustrated sufficiently by a one-dimensional harmonic oscillator describing Brownian motion. The appropriate stochastic equation is

$$du/dt=-\beta u+A(t)-\omega^2 x, \tag{186}$$

where ω denotes the circular frequency of the oscillator. We can write Eq. (184) alternatively in the form

$$d^2x/dt^2+\beta dx/dt+\omega^2 x=A(t). \tag{187}$$

What we seek from this equation are, of course, the probability distributions $W(x, t; x_0, u_0)$, $W(u, t; x_0, u_0)$ and $W(x, u, t; x_0, u_0)$. To obtain these distributions we first write down the formal solution of Eq. (187) regarded as an ordinary differential equation. The method of solution most appropriate for our present purposes is that of the variation of the parameters. In this method, as applied to Eq. (187), we express the solution in terms of that of the homogeneous equation:

$$x=a_1\exp(\mu_1 t)+a_2\exp(\mu_2 t) \tag{188}$$

where μ_1 and μ_2 are the roots of

$$\mu^2+\beta\mu+\omega^2=0; \tag{189}$$

i.e.,

$$\mu_1=-\tfrac{1}{2}\beta+(\tfrac{1}{4}\beta^2-\omega^2)^{\frac{1}{2}}; \quad \mu_2=-\tfrac{1}{2}\beta-(\tfrac{1}{4}\beta^2-\omega^2)^{\frac{1}{2}}. \tag{190}$$

We assume that the solution of Eq. (187) is of the form (188) where a_1 and a_2 are functions of time restricted however to satisfy the equation

$$\exp(\mu_1 t)(da_1/dt)+\exp(\mu_2 t)(da_2/dt)=0. \tag{191}$$

From Eq. (187) we derive the further relation

$$\mu_1 \exp(\mu_1 t)(da_1/dt) + \mu_2 \exp(\mu_2 t)(da_2/dt) = A(t). \tag{192}$$

Solving Eqs. (191) and (192) we readily obtain the integrals

$$\left.\begin{aligned} a_1 &= +\frac{1}{\mu_1-\mu_2}\int_0^t \exp(-\mu_1\xi)A(\xi)d\xi + a_{10}, \\ a_2 &= -\frac{1}{\mu_1-\mu_2}\int_0^t \exp(-\mu_2\xi)A(\xi)d\xi + a_{20}, \end{aligned}\right\} \tag{193}$$

where a_{10} and a_{20} are constants. Accordingly, we have the solution

$$x = \frac{1}{\mu_1-\mu_2}\left\{\exp(\mu_1 t)\int_0^t \exp(-\mu_1\xi)A(\xi)d\xi - \exp(\mu_2 t)\int_0^t \exp(-\mu_2\xi)A(\xi)d\xi\right\}$$
$$+a_{10}\exp(\mu_1 t) + a_{20}\exp(\mu_2 t). \tag{194}$$

From the foregoing equation we obtain for the velocity u the formula

$$u = \frac{1}{\mu_1-\mu_2}\left\{\mu_1\exp(\mu_1 t)\int_0^t \exp(-\mu_1\xi)A(\xi)d\xi - \mu_2\exp(\mu_2 t)\int_0^t \exp(-\mu_2\xi)A(\xi)d\xi\right\}$$
$$+\mu_1 a_{10}\exp(\mu_1 t) + \mu_2 a_{20}\exp(\mu_2 t). \tag{195}$$

The constants a_{10} and a_{20} can now be determined from the conditions that $x = x_0$ and $u = u_0$ at $t = 0$. We find

$$a_{10} = -\frac{x_0\mu_2 - u_0}{\mu_1-\mu_2}; \quad a_{20} = +\frac{x_0\mu_1 - u_0}{\mu_1-\mu_2}. \tag{196}$$

Thus, we have the solutions

$$x + \frac{1}{\mu_1-\mu_2}[(x_0\mu_2-u_0)\exp(\mu_1 t) - (x_0\mu_1-u_0)\exp(\mu_2 t)] = \int_0^t A(\xi)\psi(\xi)d\xi, \tag{197}$$

and

$$u + \frac{1}{\mu_1-\mu_2}[\mu_1(x_0\mu_2-u_0)\exp(\mu_1 t) - \mu_2(x_0\mu_1-u_0)\exp(\mu_2 t)] = \int_0^t A(\xi)\phi(\xi)d\xi, \tag{198}$$

where we have written

$$\left.\begin{aligned} \psi(\xi) &= \frac{1}{\mu_1-\mu_2}[\exp[\mu_1(t-\xi)] - \exp[\mu_2(t-\xi)]], \\ \phi(\xi) &= \frac{1}{\mu_1-\mu_2}[\mu_1\exp[\mu_1(t-\xi)] - \mu_2\exp[\mu_2(t-\xi)]]. \end{aligned}\right\} \tag{199}$$

It is now seen that the quantities on the right-hand sides of Eqs. (197) and (198) are of the forms considered in lemmas I and II in §2. Accordingly, we can at once write down the distribution functions $W(x, t; x_0, u_0)$, $W(u, t; x_0, u_0)$ and $W(x, u, t; x_0, u_0)$ in terms of the integrals

$$\int_0^t \psi^2(\xi)d\xi; \quad \int_0^t \phi^2(\xi)d\xi \quad \text{and} \quad \int_0^t \psi(\xi)\phi(\xi)d\xi. \tag{200}$$

With $\psi(\xi)$ and $\phi(\xi)$ defined as in Eqs. (199) we readily verify that

$$\int_0^t \psi^2(\xi)d\xi=\frac{1}{(\mu_1-\mu_2)^2}\left[\frac{1}{2\mu_1\mu_2}(\mu_2\exp(2\mu_1t)+\mu_1\exp(2\mu_2t))-\frac{2}{\mu_1+\mu_2}(\exp[(\mu_1+\mu_2)t]-1)-\frac{\mu_1+\mu_2}{2\mu_1\mu_2}\right], \quad (201)$$

$$\int_0^t \phi^2(\xi)d\xi=\frac{1}{(\mu_1-\mu_2)^2}\left[\tfrac{1}{2}(\mu_1\exp(2\mu_1t)+\mu_2\exp(2\mu_2t))-\frac{2\mu_1\mu_2}{\mu_1+\mu_2}(\exp[(\mu_1+\mu_2)t]-1)-\tfrac{1}{2}(\mu_1+\mu_2)\right], \quad (202)$$

and

$$\int_0^t \psi(\xi)\phi(\xi)d\xi=\frac{1}{2(\mu_1-\mu_2)^2}(\exp(\mu_1t)-\exp(\mu_2t))^2. \quad (203)$$

At this point it is convenient to introduce in the foregoing expressions the values of μ_1 and μ_2 explicitly according to Eq. (190): We find that the quantities on the left-hand sides of Eqs. (197) and (198) become, respectively,

$$x-x_0e^{-\beta t/2}\cosh\tfrac{1}{2}\beta_1t-\frac{x_0\beta+2u_0}{\beta_1}e^{-\beta t/2}\sinh\tfrac{1}{2}\beta_1t, \quad (204)$$

and

$$u-u_0e^{-\beta t/2}\cosh\tfrac{1}{2}\beta_1t+\frac{2x_0\omega^2+\beta u_0}{\beta_1}e^{-\beta t/2}\sinh\tfrac{1}{2}\beta_1t, \quad (205)$$

where we have introduced the quantity β_1 defined by

$$\beta_1=(\beta^2-4\omega^2)^{\frac{1}{2}}. \quad (206)$$

Similarly, we find

$$\int_0^t \psi^2(\xi)d\xi=\frac{1}{2\omega^2\beta}-\frac{e^{-\beta t}}{2\omega^2\beta_1^2\beta}(2\beta^2\sinh^2\tfrac{1}{2}\beta_1t+\beta\beta_1\sinh\beta_1t+\beta_1^2), \quad (207)$$

$$\int_0^t \phi^2(\xi)d\xi=\frac{1}{2\beta}-\frac{e^{-\beta t}}{2\beta_1^2\beta}(2\beta^2\sinh^2\tfrac{1}{2}\beta_1t-\beta\beta_1\sinh\beta_1t+\beta_1^2), \quad (208)$$

and

$$\int_0^t \psi(\xi)\phi(\xi)d\xi=2\beta_1^{-2}e^{-\beta t}\sinh^2\tfrac{1}{2}\beta_1t. \quad (209)$$

It is seen that all the foregoing expressions remain finite and real even when β_1 is zero or imaginary. Thus, while all the expressions remain valid as they stand in the "overdamped" case (β_1 real) the formulae appropriate for the periodic (β_1 imaginary) and the aperiodic (β_1 zero) cases can be readily written down by replacing

$$\cosh\tfrac{1}{2}\beta_1t,\ \beta_1^{-1}\sinh\tfrac{1}{2}\beta_1t \quad\text{and}\quad \beta_1^{-1}\sinh\beta_1t, \quad (210)$$

respectively, by

$$\cos\omega_1t,\quad \frac{1}{2\omega_1}\sin\omega_1t \quad\text{and}\quad \frac{1}{2\omega_1}\sin 2\omega_1t \text{ where } \omega_1=(\omega^2-\tfrac{1}{4}\beta^2)^{\frac{1}{2}} \quad (211)$$

in the periodic case, and by

$$1,\ \tfrac{1}{2}t \quad\text{and}\quad t \quad (212)$$

in the aperiodic case.

As we have already remarked, we can immediately write down the distribution functions for the quantities on the left-hand sides of the Eqs. (197) and (198) [i.e., the quantities (204) and (205)] according to lemmas I and II of §2 in terms of the integrals (207)–(209). Thus,

$$W(x,t;x_0,u_0)=\left[\frac{m}{4\pi\beta kT\int_0^t\psi^2(\xi)d\xi}\right]^{\frac{1}{2}}\exp-\frac{\left(x-x_0e^{-\beta t/2}\left[\cosh\tfrac{1}{2}\beta_1t+\dfrac{\beta}{\beta_1}\sinh\tfrac{1}{2}\beta_1t\right]-\dfrac{2u_0}{\beta_1}e^{-\beta t/2}\sinh\tfrac{1}{2}\beta_1t\right)^2}{\dfrac{2kT}{m\omega^2}\left\{1-e^{-\beta t}\left(\dfrac{2\beta^2}{\beta_1^2}\sinh^2\tfrac{1}{2}\beta_1t+\dfrac{\beta}{\beta_1}\sinh\beta_1t+1\right)\right\}}. \tag{213}$$

We have similar expressions for $W(u,t;x_0,u_0)$ and $W(x,u,t;x_0,u_0)$.

The quantities of greatest interest are the moments $\langle x\rangle_{\text{Av}}$, $\langle u\rangle_{\text{Av}}$, $\langle x^2\rangle_{\text{Av}}$, $\langle u^2\rangle_{\text{Av}}$ and $\langle xu\rangle_{\text{Av}}$. We find

$$\left.\begin{aligned}
\langle x\rangle_{\text{Av}}&=x_0e^{-\beta t/2}\left(\cosh\tfrac{1}{2}\beta_1t+\frac{\beta}{\beta_1}\sinh\tfrac{1}{2}\beta_1t\right)+\frac{2u_0}{\beta_1}e^{-\beta t/2}\sinh\tfrac{1}{2}\beta_1t,\\
\langle u\rangle_{\text{Av}}&=u_0e^{-\beta t/2}\left(\cosh\tfrac{1}{2}\beta_1t-\frac{\beta}{\beta_1}\sinh\tfrac{1}{2}\beta_1t\right)-\frac{2x_0\omega^2}{\beta_1}e^{-\beta t/2}\sinh\tfrac{1}{2}\beta_1t,\\
\langle x^2\rangle_{\text{Av}}&=\langle x\rangle_{\text{Av}}^2+\frac{kT}{m\omega^2}\left\{1-e^{-\beta t}\left(2\frac{\beta^2}{\beta_1^2}\sinh^2\tfrac{1}{2}\beta_1t+\frac{\beta}{\beta_1}\sinh\beta_1t+1\right)\right\},\\
\langle u^2\rangle_{\text{Av}}&=\langle u\rangle_{\text{Av}}^2+\frac{kT}{m}\left\{1-e^{-\beta t}\left(2\frac{\beta^2}{\beta_1^2}\sinh^2\tfrac{1}{2}\beta_1t-\frac{\beta}{\beta_1}\sinh\beta_1t+1\right)\right\},\\
\langle xu\rangle_{\text{Av}}&=\langle x\rangle_{\text{Av}}\langle u\rangle_{\text{Av}}+\frac{4\beta kT}{\beta_1^2m}e^{-\beta t}\sinh^2\tfrac{1}{2}\beta_1t.
\end{aligned}\right\} \tag{214}$$

The foregoing expressions are the average values of the various quantities at time t for assigned values of x and u (namely, x_0 and u_0) at time $t=0$. We see that

$$\left.\begin{aligned}
&\langle x\rangle_{\text{Av}}\to0;\quad\langle u\rangle_{\text{Av}}\to0;\quad\langle xu\rangle_{\text{Av}}\to0,\\
&\langle x^2\rangle_{\text{Av}}\to kT/m\omega^2;\quad\langle u^2\rangle_{\text{Av}}=kT/m,
\end{aligned}\right\}\quad t\to\infty. \tag{215}$$

By averaging the various moments over all values of u_0 and remembering that

$$\langle u_0\rangle_{\text{Av}}=0;\quad\langle u_0^2\rangle_{\text{Av}}=kT/m. \tag{216}$$

we obtain from Eqs. (214) that

$$\left.\begin{aligned}
\langle\langle x\rangle\rangle_{\text{Av}}&=x_0e^{-\beta t/2}\left(\cosh\tfrac{1}{2}\beta_1t+\frac{\beta}{\beta_1}\sinh\tfrac{1}{2}\beta_1t\right),\\
\langle\langle u\rangle\rangle_{\text{Av}}&=-\frac{2x_0\omega^2}{\beta_1}e^{-\beta t/2}\sinh\tfrac{1}{2}\beta_1t,\\
\langle\langle x^2\rangle\rangle_{\text{Av}}&=\frac{kT}{m\omega^2}+\left(x_0^2-\frac{kT}{m\omega^2}\right)e^{-\beta t}\left(\cosh\tfrac{1}{2}\beta_1t+\frac{\beta}{\beta_1}\sinh\tfrac{1}{2}\beta_1t\right)^2,\\
\langle\langle u^2\rangle\rangle_{\text{Av}}&=\frac{kT}{m}+\frac{4\omega}{\beta_1^2}\left(x_0^2-\frac{kT}{m\omega^2}\right)e^{-\beta t}\sinh^2\tfrac{1}{2}\beta_1t,\\
\langle\langle xu\rangle\rangle_{\text{Av}}&=\frac{2\omega^2}{\beta_1}\left(\frac{kT}{m\omega^2}-x_0^2\right)e^{-\beta t}\sinh\tfrac{1}{2}\beta_1t\left(\cosh\tfrac{1}{2}\beta_1t+\frac{\beta}{\beta_1}\sinh\tfrac{1}{2}\beta_1t\right).
\end{aligned}\right\} \tag{217}$$

Equations (214) and (217) show how the equipartition values (215) are reached as $t\to\infty$.

4. The Fokker-Planck Equation. The Generalization of Liouville's Theorem

As we have already remarked on several occasions, in an analysis of the Brownian movement we regard as impracticable a detailed description of the motions of the individual particles. Instead, we emphasize the essential stochastic nature of the phenomenon and seek a description in terms of the probability distributions of position and/or velocity at a later time starting from given initial distributions. Thus, in our discussion of the Brownian movement of a free particle in §2 we obtain explicitly the distribution functions $W(\boldsymbol{u}, t; \boldsymbol{u}_0)$, $W(\boldsymbol{r}, t; \boldsymbol{u}_0, \boldsymbol{r}_0)$ and $W(\boldsymbol{r}, \boldsymbol{u}, t; \boldsymbol{r}_0, \boldsymbol{u}_0)$ for given initial values of $\boldsymbol{r}_0$ and $\boldsymbol{u}_0$; similarly, in §3 we determined the distributions $W(u, t; x_0, u_0)$, $W(x, t; x_0, u_0)$ and $W(x, u, t; x_0, u_0)$ for a harmonically bound particle describing Brownian motion. In deriving these distributions in §§2 and 3 we started with the Langevin equation [Eq. (132) in the field-free case, and Eq. (184) when an external field is present] and solved it in a manner appropriate to the problem. We shall now consider the question whether we cannot reduce the determination of these distribution functions to appropriate boundary value problems of suitably chosen partial differential equations. We have in mind a reduction similar to that achieved in Chapter I, §5 where we showed how, under certain circumstances, the solution to the problem of random flights can be obtained as solutions of boundary-value problems long familiar in the theory of diffusion or conduction of heat. That a similar reduction should be possible under our present circumstances is apparent when we recall that the interpretation of the problem of random flights as one in diffusion (or heat conduction) is possible only if there exist time intervals Δt long enough for the particle to suffer a large number of individual displacements but still short enough for the *net* mean square displacement $\langle|\Delta \boldsymbol{R}|^2\rangle_{\text{Av}}$ to be small and of $O(\Delta t)$. And, it is in the essence of Brownian motion that there exist time intervals Δt during which the physical parameters (like position and velocity of the Brownian particle) change by "infinitesimal" amounts while there occur a very large number of fluctuations characteristic of the motion and arising from the collisions with the molecules of the surrounding fluid.

It is clear that for the solutions of the most general problem we shall require the density function $W(\boldsymbol{r}, \boldsymbol{u}, t)$; in other words, we should really consider the problem in the six-dimensional phase space. Accordingly, we may state our principal objective by the remark that what we are seeking is essentially a generalization of Liouville's theorem of classical dynamics to include Brownian motion. But before we proceed to establish such a general theorem it will be instructive to consider the simplest problem of the Brownian motion of a free particle in the velocity space and obtain a differential equation for $W(\boldsymbol{u}, t)$; this leads us to the discussion of the Fokker-Planck equation in its most familiar form.

(i) *The Fokker-Planck Equation in Velocity Space to Describe the Brownian Motion of a Free Particle*

Let Δt denote an interval of time long compared to the periods of fluctuations of the acceleration $\boldsymbol{A}(t)$ occurring in the Langevin equation but short compared to intervals during which the velocity of a Brownian particle changes by appreciable amounts. Under these circumstances we should expect to derive the distribution function $W(\boldsymbol{u}, t+\Delta t)$ governing the probability of occurrence of $\boldsymbol{u}$ at time $t+\Delta t$ from the distribution $W(\boldsymbol{u}, t)$ at time t and a knowledge of the *transition probability* $\psi(\boldsymbol{u}; \Delta \boldsymbol{u})$ that $\boldsymbol{u}$ suffers an increment $\Delta \boldsymbol{u}$ in time Δt. More particularly, we expect the relation

$$W(\boldsymbol{u}, t+\Delta t) = \int W(\boldsymbol{u}-\Delta \boldsymbol{u}, t)\psi(\boldsymbol{u}-\Delta \boldsymbol{u}; \Delta \boldsymbol{u})d(\Delta \boldsymbol{u}), \tag{218}$$

to be valid. We may parenthetically remark that in expecting this integral equation between $W(\boldsymbol{u}, t+\Delta t)$ and $W(\boldsymbol{u}, t)$ to be true we are actually supposing that the course which a Brownian particle will take depends only on the instantaneous values of its physical parameters and is entirely independent of its whole previous history. In general probability theory, a stochastic process which has this characteristic, namely, that what happens at a given instant of time t depends *only* on the state of the system at time t is said to be a *Markoff* process. We may describe a Markoff process picturesquely by the statement that it represents "the gradual unfolding of a transition probability" in exactly the same sense as the development of a conservative dynamical system can be described as "the gradual unfolding of a contact transformation." That we should be able to idealize Brownian motion as a Markoff process appears very reasonable. But we should be careful not to conclude too hastily that every stochastic process is necessarily of the Markoff type. For, it can happen that the future course of a system is conditioned by its past history: i.e., what happens at a given instant of time t may depend on what has already happened during all time preceding t.

Returning to Eq. (218), for the case under discussion we have

$$\psi(\mathbf{u};\Delta\mathbf{u})=\frac{1}{(4\pi q\Delta t)^{\frac{3}{2}}}\exp\left(-|\Delta\mathbf{u}+\beta\mathbf{u}\Delta t|^2/4q\Delta t\right)\quad (q=\beta kT/m). \tag{219}$$

For, according to the Langevin equation [cf. Eq. (142)]

$$\Delta\mathbf{u}=-\beta\mathbf{u}\Delta t+\mathbf{B}(\Delta t) \tag{220}$$

where $\mathbf{B}(\Delta t)$ denotes the net acceleration arising from fluctuations which a Brownian particle suffers in time Δt; and, since the distribution of $\mathbf{B}(\Delta t)$ is given by Eq. (144), the transition probability (218) follows at once.

Expanding $W(\mathbf{u},t+\Delta t)$, $W(\mathbf{u}-\Delta\mathbf{u},t)$ and $\psi(\mathbf{u}-\Delta\mathbf{u};\Delta\mathbf{u})$ in Eq. (218) in the form of Taylor series, we obtain

$$\begin{aligned}W(\mathbf{u},t)&+\frac{\partial W}{\partial t}\Delta t+O(\Delta t^2)\\ &=\int_{-\infty}^{+\infty}\int_{-\infty}^{+\infty}\int_{-\infty}^{+\infty}\left\{W(\mathbf{u},t)-\sum_i\frac{\partial W}{\partial u_i}\Delta u_i+\frac{1}{2}\sum_i\frac{\partial^2 W}{\partial u_i^2}\Delta u_i^2+\sum_{i<j}\frac{\partial^2 W}{\partial u_i\partial u_j}\Delta u_i\Delta u_j+\cdots\right\}\\ &\times\left\{\psi(\mathbf{u};\Delta\mathbf{u})-\sum_i\frac{\partial\psi}{\partial u_i}\Delta u_i+\frac{1}{2}\sum_i\frac{\partial^2\psi}{\partial u_i^2}\Delta u_i^2+\sum_{i<j}\frac{\partial^2\psi}{\partial u_i\partial u_j}\Delta u_i\Delta u_j+\cdots\right\}d(\Delta u_1)d(\Delta u_2)d(\Delta u_3)\end{aligned} \tag{221}$$

or, writing

$$\left.\begin{aligned}\langle\Delta u_i\rangle_{\mathrm{Av}}&=\int_{-\infty}^{+\infty}\Delta u_i\psi(\mathbf{u};\Delta\mathbf{u})d(\Delta\mathbf{u}),\\ \langle\Delta u_i^2\rangle_{\mathrm{Av}}&=\int_{-\infty}^{+\infty}\Delta u_i^2\psi(\mathbf{u};\Delta\mathbf{u})d(\Delta\mathbf{u}),\\ \langle\Delta u_i\Delta u_j\rangle_{\mathrm{Av}}&=\int_{-\infty}^{+\infty}\Delta u_i\Delta u_j\psi(\mathbf{u};\Delta\mathbf{u})d(\Delta\mathbf{u}),\end{aligned}\right\} \tag{222}$$

we have

$$\begin{aligned}\frac{\partial W}{\partial t}\Delta t+O(\Delta t^2)=&-\sum_i\frac{\partial W}{\partial u_i}\langle\Delta u_i\rangle_{\mathrm{Av}}+\frac{1}{2}\sum_i\frac{\partial^2 W}{\partial u_i^2}\langle\Delta u_i^2\rangle_{\mathrm{Av}}+\sum_{i<j}\frac{\partial^2 W}{\partial u_i\partial u_j}\langle\Delta u_i\Delta u_j\rangle_{\mathrm{Av}}-\sum_i W\frac{\partial}{\partial u_i}\langle\Delta u_i\rangle_{\mathrm{Av}}\\ &+\sum_i\frac{\partial}{\partial u_i}\langle\Delta u_i^2\rangle_{\mathrm{Av}}\frac{\partial W}{\partial u_i}+\sum_{i\neq j}\frac{\partial W}{\partial u_i}\frac{\partial}{\partial u_j}\langle\Delta u_i\Delta u_j\rangle_{\mathrm{Av}}+\frac{1}{2}\sum_i\frac{\partial^2}{\partial u_i^2}\langle\Delta u_i^2\rangle_{\mathrm{Av}}W\\ &+\sum_{i<j}W\frac{\partial^2}{\partial u_i\partial u_j}\langle\Delta u_i\Delta u_j\rangle_{\mathrm{Av}}+O(\langle\Delta u_i\Delta u_j\Delta u_k\rangle_{\mathrm{Av}}),\end{aligned} \tag{223}$$

where the remainder term involves the averages of the quantities

$$\Delta u_i^3, \quad \Delta u_i^2\Delta u_j \quad \text{and} \quad \Delta u_i\Delta u_j\Delta u_k, \quad (i,\, j,\, k=1,\, 2,\, 3).$$

Equation (223) can be written more conveniently as

$$\frac{\partial W}{\partial t}\Delta t+O(\Delta t^2)=-\sum_i \frac{\partial}{\partial u_i}(W\langle\Delta u_i\rangle_{\text{Av}})+\frac{1}{2}\sum_i \frac{\partial^2}{\partial u_i^2}(W\langle\Delta u_i^2\rangle_{\text{Av}})+\sum_{i<j}\frac{\partial^2}{\partial u_i\partial u_j}(W\langle\Delta u_i\Delta u_j\rangle_{\text{Av}})+O(\langle\Delta u_i\Delta u_j\Delta u_k\rangle_{\text{Av}}), \quad (224)$$

which is the *Fokker-Planck equation* in its most general form.

For the transition probability (219),

$$\langle\Delta u_i\rangle_{\text{Av}}=-\beta u_i\Delta t; \quad \langle\Delta u_i\Delta u_j\rangle_{\text{Av}}=O(\Delta t^2); \quad \langle\Delta u_i^2\rangle_{\text{Av}}=2q\Delta t+O(\Delta t^2). \quad (225)$$

Hence, Eq. (224) reduces in our case to

$$\frac{\partial W}{\partial t}\Delta t+O(\Delta t^2)=\{\beta\, \text{div}_{\boldsymbol{u}}\,(W\boldsymbol{u})+q\nabla_{\boldsymbol{u}}^2W\}\Delta t+O(\Delta t^2), \quad (226)$$

and passing now to the limit $\Delta t=0$ we have

$$\partial W/\partial t=\beta\, \text{div}_{\boldsymbol{u}}\,(W\boldsymbol{u})+q\nabla_{\boldsymbol{u}}^2W. \quad (227)$$

We shall now show that the distribution function $W(\boldsymbol{u},\, t;\, \boldsymbol{u}_0)$ obtained in §2, Eq. (161) is the fundamental solution of the Fokker-Planck Eq. (227) in the sense that this is the solution which tends to the δ function

$$\delta(u_1-u_{1,\,0})\delta(u_2-u_{2,\,0})\delta(u_3-u_{3,\,0}) \quad (228)$$

as $t\rightarrow 0$. To prove this, we first note that but for the Laplacian term, Eq. (227) is a linear partial differential equation of the first order. Hence, it is natural to expect that the general solution of Eq. (227) will be intimately connected with that of the associated first-order equation

$$(\partial W/\partial t)-\beta\, \text{div}_{\boldsymbol{u}}\,(W\boldsymbol{u})=0. \quad (229)$$

The general solution of this first-order equation involves the three first integrals of the Lagrangian subsidiary system

$$d\boldsymbol{u}/dt=-\beta\boldsymbol{u}. \quad (230)$$

The required first integrals are therefore

$$\boldsymbol{u}e^{\beta t}=\boldsymbol{u}_0=\text{constant}. \quad (231)$$

Accordingly, for solving Eq. (227) we introduce a new vector $\boldsymbol{\varrho}$ defined by

$$\boldsymbol{\varrho}=(\xi,\, \eta,\, \zeta)=\boldsymbol{u}e^{\beta t}. \quad (232)$$

Equation (227) now becomes

$$\partial W/\partial t=3\beta W+qe^{2\beta t}\nabla_{\boldsymbol{\varrho}}^2W. \quad (233)$$

This equation can be further simplified by introducing the variable

$$\chi=We^{-3\beta t}. \quad (234)$$

We have

$$\frac{\partial\chi}{\partial t}=qe^{2\beta t}\left(\frac{\partial^2\chi}{\partial\xi^2}+\frac{\partial^2\chi}{\partial\eta^2}+\frac{\partial^2\chi}{\partial\zeta^2}\right). \quad (235)$$

The solution of this equation can be readily written down by using the following lemma:

Lemma I. If $\phi(t)$ is an arbitrary function of time, the solution of the partial differential equation

$$\partial\chi/\partial t=\phi^2(t)\nabla_{\varrho}^2\chi \tag{236}$$

which has a source at $\varrho=\varrho_0$ at time $t=0$ is

$$\chi=\frac{1}{\left[4\pi\int_0^t\phi^2(t)dt\right]^{\frac{3}{2}}}\exp\left(-|\varrho-\varrho_0|^2\bigg/4\int_0^t\phi^2(t)dt\right). \tag{237}$$

We shall omit the proof of this lemma as it is very elementary.

Applying this lemma to Eq. (235) we have the fundamental solution

$$\chi=\frac{1}{\left[4\pi q\int_0^t e^{2\beta t}dt\right]^{\frac{3}{2}}}\exp\left(-|\boldsymbol{u}e^{\beta t}-\boldsymbol{u}_0|^2\bigg/4q\int_0^t e^{2\beta t}dt\right), \tag{238}$$

or, returning to the variable W according to Eq. (234) we have

$$W(\boldsymbol{u},t;\boldsymbol{u}_0)=\frac{1}{[2\pi q(1-e^{-2\beta t})/\beta]^{\frac{3}{2}}}\exp\left[-\beta|\boldsymbol{u}-\boldsymbol{u}_0e^{-\beta t}|^2/2q(1-e^{-2\beta t})\right] \tag{239}$$

which agrees with our earlier result in §2, Eq. (161).

(*ii*) *The Generalization of Liouville's Theorem to Include Brownian Motion*

We shall now consider the general problem of a particle describing Brownian motion and under the influence of an external field of force.

Let Δt again denote an interval of time which is long compared to the periods of fluctuations of the acceleration $\boldsymbol{A}(t)$ occurring in the Langevin Eq. (184) but short compared to the intervals in which any of the physical parameters change appreciably. Then, the increments $\Delta\boldsymbol{r}$ and $\Delta\boldsymbol{u}$ in position and velocity which the particle suffers during Δt are

$$\Delta\boldsymbol{r}=\boldsymbol{u}\Delta t;\quad \Delta\boldsymbol{u}=-(\beta\boldsymbol{u}-\boldsymbol{K})\Delta t+\boldsymbol{B}(\Delta t), \tag{240}$$

where $\boldsymbol{K}$ denotes the acceleration per unit mass caused by the external field of force and $\boldsymbol{B}(\Delta t)$ the net acceleration arising from fluctuations which the particle suffers in time Δt. The distribution of $\boldsymbol{B}(\Delta t)$ is again given by Eq. (144).

Assuming as before that the Brownian movement can be idealized as a Markoff process the probability distribution $W(\boldsymbol{r},\boldsymbol{u},t+\Delta t)$ in *phase space* at time $t+\Delta t$ can be derived from the distribution $W(\boldsymbol{r},\boldsymbol{u},t)$ at the earlier time t by means of the integral equation

$$W(\boldsymbol{r},\boldsymbol{u},t+\Delta t)=\int\int W(\boldsymbol{r}-\Delta\boldsymbol{r},\boldsymbol{u}-\Delta\boldsymbol{u},t)\Psi(\boldsymbol{r}-\Delta\boldsymbol{r},\boldsymbol{u}-\Delta\boldsymbol{u};\Delta\boldsymbol{r},\Delta\boldsymbol{u})d(\Delta\boldsymbol{r})d(\Delta\boldsymbol{u}). \tag{241}$$

According to the Eqs. (240) we can write

$$\Psi(\boldsymbol{r},\boldsymbol{u};\Delta\boldsymbol{r},\Delta\boldsymbol{u})=\psi(\boldsymbol{r},\boldsymbol{u};\Delta\boldsymbol{u})\delta(\Delta x-u_1\Delta t)\delta(\Delta y-u_2\Delta t)\delta(\Delta z-u_3\Delta t), \tag{242}$$

where the δ's denote Dirac's δ functions and $\psi(\boldsymbol{r},\boldsymbol{u};\Delta\boldsymbol{u})$ the transition probability in the velocity space. With this form for the transition probability in the phase space the integration over $\Delta\boldsymbol{r}$ in

Eq. (241) is immediately performed and we get

$$W(\mathbf{r}, \mathbf{u}, t+\Delta t)=\int W(\mathbf{r}-\mathbf{u}\Delta t, \mathbf{u}-\Delta \mathbf{u}, t)\psi(\mathbf{r}-\mathbf{u}\Delta t, \mathbf{u}-\Delta \mathbf{u}; \Delta \mathbf{u})d(\Delta \mathbf{u}). \tag{243}$$

Alternatively, we can write

$$W(\mathbf{r}+\mathbf{u}\Delta t, \mathbf{u}, t+\Delta t)=\int W(\mathbf{r}, \mathbf{u}-\Delta \mathbf{u}, \Delta t)\psi(\mathbf{r}, \mathbf{u}-\Delta \mathbf{u}; \Delta \mathbf{u})d(\Delta \mathbf{u}). \tag{244}$$

Expanding the various functions in the foregoing equation in the form of Taylor series and proceeding as in our derivation of the Fokker-Planck equation, we obtain [cf. Eq. (221)]

$$\left(\frac{\partial W}{\partial t}+\mathbf{u}\cdot\text{grad}_r\, W\right)\Delta t+O(\Delta t^2)=-\sum_i \frac{\partial}{\partial u_i}(W\langle\Delta u_i\rangle_{\text{Av}})+\frac{1}{2}\sum_i \frac{\partial^2}{\partial u_i^2}(W\langle\Delta u_i^2\rangle_{\text{Av}})$$
$$+\sum_{i<j}\frac{\partial^2}{\partial u_i \partial u_j}(W\langle\Delta u_i\Delta u_j\rangle_{\text{Av}})+O(\langle\Delta u_i\Delta u_j\Delta u_k\rangle_{\text{Av}}). \tag{245}$$

This is the complete analog in the phase space of the Fokker-Planck equation in the velocity space.

For the case (240), the transition probability $\psi(\mathbf{u}; \Delta\mathbf{u})$ is given by [cf. Eq. (144)]

$$\psi(\mathbf{u}; \Delta\mathbf{u})=\frac{1}{(4\pi q\Delta t)^{\frac{3}{2}}}\exp\left(-|\Delta\mathbf{u}+(\beta\mathbf{u}-\mathbf{K})\Delta t|^2/4q\Delta t\right). \tag{246}$$

And with this expression for the transition probability we clearly have

$$\langle\Delta u_i\rangle_{\text{Av}}=-(\beta u_i-K_i)\Delta t;\quad \langle\Delta u_i^2\rangle_{\text{Av}}=2q\Delta t+O(\Delta t^2);\quad \langle\Delta u_i\Delta u_j\rangle_{\text{Av}}=O(\Delta t^2). \tag{247}$$

Accordingly Eq. (245) simplifies to

$$\left\{\frac{\partial W}{\partial t}+\mathbf{u}\cdot\text{grad}_r\, W\right\}\Delta t+O(\Delta t^2)=\left\{\sum_i \frac{\partial}{\partial u_i}[(\beta u_i-K_i)W]+q\sum_i\frac{\partial^2 W}{\partial u_i^2}\right\}\Delta t+O(\Delta t^2), \tag{248}$$

and now passing to the limit $\Delta t=0$ and after some minor rearranging of the terms we finally obtain

$$\partial W/\partial t+\mathbf{u}\cdot\text{grad}_r\, W+\mathbf{K}\cdot\text{grad}_u\, W=\beta\,\text{div}_u\,(W\mathbf{u})+q\nabla_u^2 W. \tag{249}$$

The foregoing equation represents the complete generalization of the Fokker-Planck Eq. (227) to the phase space. At the same time Eq. (249) represents also the generalization of Liouville's theorem of classical dynamics to include Brownian motion; more particularly, on the right-hand side of Eq. (249), we have the terms arising from Brownian motion while on the left-hand side we have the usual Stokes operator D/Dt acting on W.

(*iii*) *The Solution of Equation* (249) *for the Field Free Case*

When no external field is present Eq. (249) becomes

$$\partial W/\partial t+\mathbf{u}\cdot\text{grad}_r\, W=3\beta W+\beta\mathbf{u}\cdot\text{grad}_u\, W+q\nabla_u^2 W. \tag{250}$$

To solve this equation we again note that the equation

$$\partial W/\partial t+\mathbf{u}\cdot\text{grad}_r\, W=3\beta W+\beta\mathbf{u}\cdot\text{grad}_u\, W \tag{251}$$

derived from (250) by ignoring the Laplacian term $q\nabla_u^2 W$ is a linear homogeneous first-order partial differential equation for $We^{-3\beta t}$. Accordingly, the general solution of Eq. (251) can be expressed in terms of any six independent integrals of the Lagrangian subsidiary system

$$du/dt=-\beta \mathbf{u};\quad d\mathbf{r}/dt=\mathbf{u}. \tag{252}$$

Two vector integrals of this system are

$$\mathbf{u}e^{\beta t}=\mathbf{I}_1;\quad \mathbf{r}+\mathbf{u}/\beta=\mathbf{I}_2. \tag{253}$$

Accordingly, to solve Eq. (119) we introduce the new variables

$$\boldsymbol{\varrho}=(\xi,\eta,\zeta)=\mathbf{u}e^{\beta t};\quad \mathbf{P}=(X,Y,Z)=\mathbf{r}+\mathbf{u}/\beta. \tag{254}$$

For this transformation of the variables we have

$$\left.\begin{aligned}\frac{\partial W}{\partial t}&=\frac{\partial}{\partial t}W(\boldsymbol{\varrho},\mathbf{P},t)+\beta\boldsymbol{\varrho}\cdot\mathrm{grad}_{\boldsymbol{\varrho}}\,W,\\ \mathrm{grad}_{\mathbf{r}}\,W&=\mathrm{grad}_{\mathbf{P}}\,W,\\ \mathrm{grad}_{\mathbf{u}}\,W&=e^{\beta t}\,\mathrm{grad}_{\boldsymbol{\varrho}}\,W+(1/\beta)\,\mathrm{grad}_{\mathbf{P}}\,W,\end{aligned}\right\} \tag{255}$$

and finally

$$\nabla_{\mathbf{u}}^2W=e^{2\beta t}\nabla_{\boldsymbol{\varrho}}^2W+(2/\beta)e^{\beta t}\nabla_{\boldsymbol{\varrho}}\cdot\nabla_{\mathbf{P}}W+(1/\beta^2)\nabla_{\mathbf{P}}^2W. \tag{256}$$

Substituting the foregoing equations in Eq. (250) we obtain

$$\partial W/\partial t=3\beta W+q\{e^{2\beta t}\nabla_{\boldsymbol{\varrho}}^2W+(2/\beta)e^{\beta t}\nabla_{\boldsymbol{\varrho}}\cdot\nabla_{\mathbf{P}}W+(1/\beta^2)\nabla_{\mathbf{P}}^2W\}. \tag{257}$$

Again, we introduce the variable

$$\chi=We^{-3\beta t}. \tag{258}$$

Equation (257) reduces to

$$\partial\chi/\partial t=q\{e^{2\beta t}\nabla_{\boldsymbol{\varrho}}^2\chi+(2/\beta)e^{\beta t}\nabla_{\boldsymbol{\varrho}}\cdot\nabla_{\mathbf{P}}\chi+(1/\beta^2)\nabla_{\mathbf{P}}^2\chi\}, \tag{259}$$

or, written out explicitly

$$\frac{\partial\chi}{\partial t}=q\left\{e^{2\beta t}\left(\frac{\partial^2\chi}{\partial\xi^2}+\frac{\partial^2\chi}{\partial\eta^2}+\frac{\partial^2\chi}{\partial\zeta^2}\right)+\frac{2}{\beta}e^{\beta t}\left(\frac{\partial^2\chi}{\partial\xi\partial X}+\frac{\partial^2\chi}{\partial\eta\partial Y}+\frac{\partial^2\chi}{\partial\zeta\partial Z}\right)+\frac{1}{\beta^2}\left(\frac{\partial^2\chi}{\partial X^2}+\frac{\partial^2\chi}{\partial Y^2}+\frac{\partial^2\chi}{\partial Z^2}\right)\right\}. \tag{260}$$

To solve this equation we first prove the following lemma:

Lemma II. Let $\phi(t)$ and $\psi(t)$ be two arbitrary functions of time. The solution of the differential equation

$$\frac{\partial\chi}{\partial t}=\phi^2(t)\frac{\partial^2\chi}{\partial\xi^2}+2\phi(t)\psi(t)\frac{\partial^2\chi}{\partial\xi\partial X}+\psi^2(t)\frac{\partial^2\chi}{\partial X^2} \tag{261}$$

which has a source at $\xi=X=0$ at $t=0$ is

$$\chi=\frac{1}{2\pi\Delta^{\frac{1}{2}}}\exp\left[-(a\xi^2+2h\xi X+bX^2)/2\Delta\right], \tag{262}$$

where

$$a=2\int_0^t\psi^2(t)dt;\quad h=-2\int_0^t\phi(t)\psi(t)dt;\quad b=2\int_0^t\phi^2(t)dt, \tag{263}$$

and

$$\Delta=ab-h^2. \tag{264}$$

To prove this lemma we substitute for χ according to Eq. (262) in the differential Eq. (261). After some minor reductions we find that we are left with

$$\frac{1}{\Delta}\frac{d\Delta}{dt}+\xi^2\frac{da_1}{dt}+2\xi X\frac{dh_1}{dt}+X^2\frac{db_1}{dt}+2\phi^2(a_1{}^2\xi^2+2a_1h_1\xi X+h_1{}^2X^2-a_1)$$
$$+4\phi\psi(a_1h_1\xi^2+h_1b_1X^2+\xi X[h_1{}^2+a_1b_1]-h_1)+2\psi^2(h_1{}^2\xi^2+2h_1b_1\xi X+b_1{}^2X^2-b_1)=0, \quad (265)$$

where we have written

$$a_1=a/\Delta;\quad h_1=h/\Delta;\quad b_1=b/\Delta. \quad (266)$$

Equating the coefficients of ξ^2, ξX and X^2 in (265) we obtain the set of equations

$$\left.\begin{aligned} da_1/dt&=-2(a_1\phi+h_1\psi)^2,\\ db_1/dt&=-2(h_1\phi+b_1\psi)^2,\\ dh_1/dt&=-2(a_1\phi+h_1\psi)(h_1\phi+b_1\psi),\end{aligned}\right\} \quad (267)$$

and

$$d\Delta/dt=2\Delta(a_1\phi^2+2h_1\phi\psi+b_1\psi^2). \quad (268)$$

It is readily verified that Eq. (268) is consistent with the Eq. (267) [see Eqs. (271) and (272) below].
Since [cf. Eqs. (266)]

$$da/dt=\Delta(da_1/dt)+a_1(d\Delta/dt), \quad (269)$$

we have according to Eqs. (267) and (268)

$$da/dt=-2\Delta(a_1\phi+h_1\psi)^2+2\Delta(a_1{}^2\phi^2+2a_1h_1\phi\psi+a_1b_1\psi^2)=2\Delta(a_1b_1-h_1{}^2)\psi^2, \quad (270)$$

or

$$da/dt=2\psi^2. \quad (271)$$

Similarly we prove that

$$db/dt=2\phi^2;\quad dh/dt=-2\phi\psi. \quad (272)$$

Hence,

$$a=2\int^t\psi^2dt;\quad h=-2\int^t\phi\psi dt;\quad b=2\int^t\phi^2dt. \quad (273)$$

The lemma now follows as an immediate consequence of the boundary conditions at $t=0$ stated.

In order to apply the foregoing lemma to Eq. (260) we first notice that the equation is separable in the pairs of variables (ξ, X), (η, Y) and (ζ, Z). Expressing therefore the solution in the form

$$\chi=\chi_1(\xi, X)\chi_2(\eta, Y)\chi_3(\zeta, Z), \quad (274)$$

we see that each of the functions χ_1, χ_2 and χ_3 satisfies an equation of the form (261) with

$$\phi(t)=q^{\frac{1}{2}}e^{\beta t};\quad \psi(t)=q^{\frac{1}{2}}/\beta. \quad (275)$$

Hence, the solution of Eq. (260) with the boundary condition

$$\boldsymbol{\varrho}=\boldsymbol{\varrho}_0,\quad \boldsymbol{P}=\boldsymbol{P}_0\quad\text{at}\quad t=0 \quad (276)$$

is

$$\chi=\frac{1}{8\pi^3\Delta^{\frac{3}{2}}}\exp\left\{-[a|\boldsymbol{\varrho}-\boldsymbol{\varrho}_0|^2+2h(\boldsymbol{\varrho}-\boldsymbol{\varrho}_0)\cdot(\boldsymbol{P}-\boldsymbol{P}_0)+b|\boldsymbol{P}-\boldsymbol{P}_0|^2]/2\Delta\right\} \quad (277)$$

where

$$\left.\begin{aligned} a&=2q\beta^{-2}\int_0^t dt=2q\beta^{-2}t,\\ b&=2q\int_0^t e^{2\beta t}dt=q\beta^{-1}(e^{2\beta t}-1),\\ h&=-2q\beta^{-1}\int_0^t e^{\beta t}dt=-2q\beta^{-2}(e^{\beta t}-1),\end{aligned}\right\} \quad (278)$$

and

$$\boldsymbol{\varrho}-\boldsymbol{\varrho}_0=e^{\beta t}\boldsymbol{u}-\boldsymbol{u}_0;\quad \boldsymbol{P}-\boldsymbol{P}_0=\boldsymbol{r}+\boldsymbol{u}/\beta-\boldsymbol{r}_0-\boldsymbol{u}_0/\beta. \tag{279}$$

In Eq. (279) $\boldsymbol{r}_0$ and $\boldsymbol{u}_0$ denote the position and velocity of the Brownian particle at time $t=0$. Finally,

$$W=\frac{e^{3\beta t}}{8\pi^3\Delta^{\frac{3}{2}}}\exp\{-[a|\boldsymbol{\varrho}-\boldsymbol{\varrho}_0|^2+2h(\boldsymbol{\varrho}-\boldsymbol{\varrho}_0)\cdot(\boldsymbol{P}-\boldsymbol{P}_0)+b|\boldsymbol{P}-\boldsymbol{P}_0|^2]/2\Delta\}. \tag{280}$$

We shall now verify that the foregoing solution for W obtained as the fundamental solution of Eq. (250) agrees with what we obtained in §2 through a discussion of the Langevin equation: With $\boldsymbol{R}$ and $\boldsymbol{S}$ as defined in Eqs. (181) we have

$$\boldsymbol{\varrho}-\boldsymbol{\varrho}_0=e^{\beta t}\boldsymbol{S};\quad \boldsymbol{P}-\boldsymbol{P}_0=\boldsymbol{R}+(1/\beta)\boldsymbol{S}. \tag{281}$$

Accordingly,

$$\left.\begin{aligned} a|\boldsymbol{\varrho}-\boldsymbol{\varrho}_0|^2+2h(\boldsymbol{\varrho}-\boldsymbol{\varrho}_0)\cdot(\boldsymbol{P}-\boldsymbol{P}_0)+b|\boldsymbol{P}-\boldsymbol{P}_0|^2&=ae^{2\beta t}|\boldsymbol{S}|^2+2he^{\beta t}(\boldsymbol{R}\cdot\boldsymbol{S}+(1/\beta)|\boldsymbol{S}|^2)+b|\boldsymbol{R}+(1/\beta)\boldsymbol{S}|^2,\\ &=e^{2\beta t}(F|\boldsymbol{S}|^2-2H\boldsymbol{R}\cdot\boldsymbol{S}+G|\boldsymbol{R}|^2),\end{aligned}\right\} \tag{282}$$

where

$$F=a+2h\beta^{-1}e^{-\beta t}+b\beta^{-2}e^{-2\beta t};\quad G=be^{-2\beta t};\quad H=-(he^{-\beta t}+b\beta^{-1}e^{-2\beta t}). \tag{283}$$

With a, b and h as given by Eqs. (278) we find that

$$F=q\beta^{-3}(2\beta t-3+4e^{-\beta t}-e^{-2\beta t});\quad G=q\beta^{-1}(1-e^{-2\beta t});\quad H=q\beta^{-2}(1-e^{-\beta t})^2. \tag{284}$$

Further,

$$FG-H^2=(ab-h^2)e^{-2\beta t}=\Delta e^{-2\beta t}. \tag{285}$$

Thus the solution (280) can be expressed alternatively in the form

$$W=\frac{1}{8\pi^3(FG-H^2)^{\frac{3}{2}}}\exp[-(F|\boldsymbol{S}|^2-2H\boldsymbol{R}\cdot\boldsymbol{S}+G|\boldsymbol{R}|^2)/2(FG-H^2)]. \tag{286}$$

Comparing Eqs. (284) and (286) with Eqs. (178), (182) and (183) we see that the verification is complete.

(iv) *The Solution of Equation* (249) *for the Case of a Harmonically Bound Particle*

The method of solution is sufficiently illustrated by considering the case of a one-dimensional oscillator describing Brownian motion. Equation (249) then reduces to

$$\frac{\partial W}{\partial t}+u\frac{\partial W}{\partial x}-\omega^2x\frac{\partial W}{\partial u}=\beta u\frac{\partial W}{\partial u}+\beta W+q\frac{\partial^2 W}{\partial u^2}. \tag{287}$$

As in our discussion in the two preceding sections we introduce as variables two first integrals of the associated subsidiary system:

$$dx/dt=u;\quad du/dt=-\beta u-\omega^2 x. \tag{288}$$

Two independent first integrals of Eqs. (288) are readily seen to be

$$(x\mu_1-u)\exp(-\mu_2 t)\quad\text{and}\quad(x\mu_2-u)\exp(-\mu_1 t) \tag{289}$$

where μ_1 and μ_2 have the same meanings as in §3 [cf. Eqs. (189) and (190)]. Accordingly we set

$$\xi=(x\mu_1-u)\exp(-\mu_2 t);\quad \eta=(x\mu_2-u)\exp(-\mu_1 t). \tag{290}$$

In these variables Eq (287) becomes

$$\frac{\partial W}{\partial t}=\beta W+q\left(\exp(-2\mu_2 t)\frac{\partial^2 W}{\partial \xi^2}+2\exp(-(\mu_1+\mu_2)t)\frac{\partial^2 W}{\partial\xi\partial\eta}+\exp(-2\mu_1 t)\frac{\partial^2 W}{\partial\eta^2}\right). \tag{291}$$

Introducing the further transformation

$$W=\chi e^{\beta t}, \tag{292}$$

we finally obtain

$$\frac{\partial \chi}{\partial t}=q\left(\exp(-2\mu_2 t)\frac{\partial^2 \chi}{\partial \xi^2}+2\exp[-(\mu_1+\mu_2)t]\frac{\partial^2 \chi}{\partial\xi\partial\eta}+\exp(-2\mu_1 t)\frac{\partial^2 \chi}{\partial\eta^2}\right). \tag{293}$$

This equation is of the same form as Eq. (261) in lemma II. Hence the solution of this equation which tends to $\delta(\xi-\xi_0)\delta(\eta-\eta_0)$ as $t\rightarrow 0$ is

$$\chi=\frac{1}{2\pi\Delta^{\frac{1}{2}}}\exp\{-[a(\xi-\xi_0)^2+2h(\xi-\xi_0)(\eta-\eta_0)+b(\eta-\eta_0)^2]/2\Delta\}, \tag{294}$$

where

$$\left.\begin{aligned} a&=2q\int_0^t\exp(-2\mu_1 t)dt=\frac{q}{\mu_1}[1-\exp(-2\mu_1 t)],\\ b&=2q\int_0^t\exp(-2\mu_2 t)dt=\frac{q}{\mu_2}[1-\exp(-2\mu_2 t)],\\ h&=-2q\int_0^t\exp[-(\mu_1+\mu_2)t]dt=-\frac{2q}{\mu_1+\mu_2}\{1-\exp[-(\mu_1+\mu_2)t]\}.\end{aligned}\right\} \tag{295}$$

Further,

$$\xi_0=x_0\mu_1-u_0;\quad \eta_0=x_0\mu_2-u_0, \tag{296}$$

where x_0 and u_0 denote the position and velocity of the particle at time $t=0$. It is again verified that the solution

$$W=\frac{e^{\beta t}}{2\pi\Delta^{\frac{1}{2}}}\exp\{-[a(\xi-\xi_0)^2+2h(\xi-\xi_0)(\eta-\eta_0)+b(\eta-\eta_0)^2]/2\Delta\}, \tag{297}$$

obtained as the fundamental solution of Eq. (287) is in agreement with the distributions obtained in §3 through a discussion of the Langevin equation.

(v) *The General Case*

Our discussion in the two preceding sections suggests that in dealing with Eq. (249) quite generally we may introduce as new variables six independent first integrals of the equations of motion

$$d\mathbf{r}/dt=\mathbf{u};\quad d\mathbf{u}/dt=-\beta\mathbf{u}+\mathbf{K}. \tag{298}$$

These are the Lagrangian subsidiary equations of the linear first-order equation derived from (249) after ignoring the Laplacian term $q\nabla_u^2 W$. If $I_1, \cdots, I_6$ are six such integrals, we introduce

$$I_1(\mathbf{r},\mathbf{u},t),\cdots,I_6(\mathbf{r},\mathbf{u},t) \tag{299}$$

as the new independent variables. If we further set

$$W=\chi e^{3\beta t}, \tag{300}$$

Eq. (249) will transform to

$$\partial\chi/\partial t=q[\nabla_u^2\chi]_{I_1,\cdots I_6}, \tag{301}$$

where the Laplacian of χ on the right-hand side has to be expressed in terms of the new variables $I_1, \cdots, I_6$.

We shall thus be left with a general linear partial differential equation of the second order for χ; and we seek a solution of this equation of the form

$$\chi = \frac{1}{8\pi^3 \Delta^{\frac{1}{2}}} e^{-Q/2\Delta}, \tag{302}$$

where Q stands for a general homogeneous quadratic form in the six variables $I_1, \cdots, I_6$ with coefficients which are functions of time only. Further, in Eq. (302) Δ is the determinant of the matrix formed by the coefficients of the quadratic form. In this manner we can expect to solve the general problem.

(vi) The Differential Equation for the Displacement ($t \gg \beta^{-1}$). The Smoluchowski Equation

We have seen that all the physically significant questions concerning the motion of a free Brownian particle can be answered by solving Eq. (250) with appropriate boundary conditions. However, if we are interested only in time intervals very large compared to the "*time of relaxation*" β^{-1} we can apply the method of the Fokker-Planck equation to configuration space ($\mathbf{r}$) independently of the velocity space. For, according to Eq. (172), we may say that for a free Brownian particle, the transition probability that $\mathbf{r}$ suffers an increment $\Delta\mathbf{r}$ in time $\Delta t \gg \beta^{-1}$ is given by

$$\psi(\Delta\mathbf{r}) = \frac{1}{(4\pi D\Delta t)^{\frac{3}{2}}} \exp\left(-|\Delta\mathbf{r}|^2/4D\Delta t\right), \tag{303}$$

where

$$D = q/\beta^2 = kT/m\beta. \tag{304}$$

Thus, again with the understanding that $\Delta t \gg \beta^{-1}$ we can write [cf. Eq. (218) and the remarks following it]

$$w(\mathbf{r}, t+\Delta t) = \int w(\mathbf{r}-\Delta\mathbf{r}, t)\psi(\Delta\mathbf{r})d(\Delta\mathbf{r}). \tag{305}$$

Applying now to this equation the procedure that was followed in the derivation of the Fokker-Planck equation in the velocity space we readily obtain the "diffusion equation"

$$\partial w/\partial t = D\nabla_r^2 w. \tag{306}$$

That we should be led to the diffusion equation is not surprising since Eq. (303) implies that for time intervals $\Delta t \gg \beta^{-1}$ the motion of the particle reduces to the elementary case of the problem of random flights (Chapter I, §4 case [c]) and the analysis of *I* §5 leading to Eq. (112) applies.

Equation (306) is valid for a free Brownian particle. To extend this result for the case when an external field is acting we start from Eq. (249) which is quite generally true in phase space. We first rewrite this equation in the form

$$\frac{\partial W}{\partial t} = \beta\left(\mathrm{div}_{\mathbf{u}} - \frac{1}{\beta}\,\mathrm{div}_{\mathbf{r}}\right)\left(W\mathbf{u} + \frac{q}{\beta}\,\mathrm{grad}_{\mathbf{u}}\, W - \frac{\mathbf{K}}{\beta}W + \frac{q}{\beta^2}\,\mathrm{grad}_{\mathbf{r}}\, W\right) + \mathrm{div}_{\mathbf{r}}\left(\frac{q}{\beta^2}\mathrm{grad}_{\mathbf{r}}\, W - \frac{\mathbf{K}}{\beta}W\right). \tag{307}$$

We now integrate this equation along the straight line

$$\mathbf{r} + \mathbf{u}/\beta = \text{constant} = \mathbf{r}_0, \tag{308}$$

from $\mathbf{u} = -\infty$ to $+\infty$. We obtain

$$\frac{\partial}{\partial t}\int_{\mathbf{r}+\mathbf{u}\beta^{-1}=\mathbf{r}_0} W d\mathbf{u} = \int_{\mathbf{r}+\mathbf{u}\beta^{-1}=\mathbf{r}_0} \mathrm{div}_{\mathbf{r}}\left(\frac{q}{\beta^2}\,\mathrm{grad}_{\mathbf{r}}\, W - \frac{\mathbf{K}}{\beta}W\right)d\mathbf{u}. \tag{309}$$

We shall now suppose that $\boldsymbol{K}(\boldsymbol{r})$ does not change appreciably over distances of the order of $(q/\beta^3)^{\frac{1}{2}}$. Then, starting from an arbitrary initial distribution $W(\boldsymbol{r}, \boldsymbol{u}, 0)$ at time $t=0$ we should expect that a Maxwellian distribution of the velocities will be established at all points after time intervals $\Delta t \gg \beta^{-1}$. Consequently, if we are not interested in time intervals of the order of β^{-1} we can write

$$W(\boldsymbol{r}, \boldsymbol{u}, t) \simeq \left(\frac{m}{2\pi kT}\right)^{\frac{3}{2}} \exp\left(-m|\boldsymbol{u}|^2/2kT\right) w(\boldsymbol{r}, t). \tag{310}$$

With these assumptions Eq. (309) becomes

$$\frac{\partial w}{\partial t} \simeq \operatorname{div}_{\boldsymbol{r}_0} \left\{ \frac{q}{\beta^2} \operatorname{grad}_{\boldsymbol{r}_0} w(\boldsymbol{r}_0) - \frac{\boldsymbol{K}(\boldsymbol{r}_0)}{\beta} w(\boldsymbol{r}_0) \right\}. \tag{311}$$

The passage from Eqs. (309) to (311) is the result of our supposition that in the domain of $\boldsymbol{u}$ from which the dominant contribution to the integral on the right-hand side of Eq. (309) arises (namely, $|\boldsymbol{u}| \lesssim (kT/m)^{\frac{1}{2}} = (q/\beta)^{\frac{1}{2}}$) the variation of $\boldsymbol{r}$ (which is of the order $|\boldsymbol{u}|/\beta \simeq (q/\beta^3)^{\frac{1}{2}}$) is small compared to the distances in the configuration space in which $\boldsymbol{K}$ and w change appreciably. The required generalization of Eq. (306) is therefore

$$\frac{\partial w}{\partial t} = \operatorname{div}_{\boldsymbol{r}} \left(\frac{q}{\beta^2} \operatorname{grad}_{\boldsymbol{r}} w - \frac{\boldsymbol{K}}{\beta} w \right). \tag{312}$$

Equation (312) is sometimes called Smoluchowski's equation.

An immediate consequence of Eq. (312) may be noted. According to this equation a *stationary diffusion current* j obeys the law

$$j = \beta^{-1}\boldsymbol{K}w - q\beta^{-2} \operatorname{grad} w = \text{constant}. \tag{313}$$

If $\boldsymbol{K}$ can be derived from a potential $\mathfrak{B}$ so that

$$\boldsymbol{K} = -\operatorname{grad} \mathfrak{B} \tag{314}$$

Eq. (313) can be rewritten in the form

$$j = -q\beta^{-2} \exp\left(-\beta\mathfrak{B}/q\right) \operatorname{grad}\left(w \exp\left(\beta\mathfrak{B}/q\right)\right), \tag{315}$$

where it may be noted $q/\beta = kT/m$. Integrating Eq. (315) between any two points A and B we obtain

$$j \cdot \int_A^B \beta \exp\left(\beta\mathfrak{B}/q\right) d\boldsymbol{s} = \frac{kT}{m} w \exp\left(\beta\mathfrak{B}/q\right) \Big|_B^A, \tag{316}$$

an important equation, first derived by Kramers.

We may finally again draw attention to the fact that Eqs. (306) and (312) are valid only if we ignore effects which happen in time intervals of the order of β^{-1} and space intervals of the order of $(q/\beta^3)^{\frac{1}{2}}$; when such effects are of interest we should go back to Eqs. (249) or (250) which are rigorously valid in phase space.

(vii) *General Remarks*

So far we have only shown that the discussion based on Eq. (249) and its various special forms leads to results in agreement with those already derived on the basis of the Langevin equation. However, the special importance of the partial differential equations arises when further restrictions on the problem are imposed. For, these additional restrictions can also be expressed in the form of boundary conditions which the solutions will have to satisfy and the consequent reduction to a boundary value problem in partial differential equations provides a very direct method for obtaining the necessary solutions. The alternative analysis based on the Langevin equation would in general be too involved.

Further examples of the use of the partial differential equations obtained in this section will be found in Chapters III and IV.

5. General Remarks

A general characteristic of the stochastic processes of the type considered in the preceding sections is that the increment in the velocity, $\Delta \boldsymbol{u}$ which a particle suffers in a time Δt long compared to the periods of the elementary fluctuations can be expressed as the sum of two distinct terms: a term $\boldsymbol{K}\Delta t$ which represents the action of the external field of force, and a term $\delta \boldsymbol{u}(\Delta t)$ which denotes a fluctuating quantity with a definite law of distribution. Thus

$$\Delta \boldsymbol{u} = \boldsymbol{K}\Delta t + \delta \boldsymbol{u}(\Delta t); \tag{317}$$

the corresponding increment in the position, $\Delta \boldsymbol{r}$ is given by

$$\Delta \boldsymbol{r} = \boldsymbol{u}\Delta t, \tag{318}$$

where $\boldsymbol{u}$ is the instantaneous velocity of the particle.

When dealing with stochastic processes of the *strictly* Brownian motion type we further suppose that the term $\delta \boldsymbol{u}(\Delta t)$ in Eq. (317) can in turn be decomposed into two parts: a part $-\beta \boldsymbol{u}\Delta t$ representing the deceleration caused by the dynamical friction $-\beta \boldsymbol{u}$ and a fluctuating part $\boldsymbol{B}(\Delta t)$ which is really the vector sum of a very large number of very "minute" accelerations arising from collisions with individual molecules of the surrounding fluid:

$$\delta \boldsymbol{u}(\Delta t) = -\beta \boldsymbol{u}\Delta t + \boldsymbol{B}(\Delta t). \tag{319}$$

It is *this* particular decomposition of $\delta \boldsymbol{u}(\Delta t)$ that is peculiarly characteristic of stochastic processes of the Brownian type.

Concerning $\boldsymbol{B}(\Delta t)$ in Eq. (319) we have supposed in §§2, 3, and 4 that it is governed by the distribution function [cf. Eqs. (144) and (145)]

$$w(\boldsymbol{B}[\Delta t]) = \frac{1}{(4\pi q\Delta t)^{\frac{3}{2}}} \exp\left(-|\boldsymbol{B}(\Delta t)|^2/4q\Delta t\right), \tag{320}$$

where

$$q = \beta kT/m. \tag{321}$$

In this choice of the distribution function for $\boldsymbol{B}(\Delta t)$ we were guided by two considerations: *First*, that starting from any arbitrarily assigned distribution of the velocities we shall always be led to the Maxwellian distribution as $t \to \infty$ (or, alternatively that the Maxwellian distribution of the velocities is invariant to stochastic processes of the type considered); and *second* that during a time Δt in which the position and the velocity of the particle will change by an "infinitesimal" amount of order Δt the particle will in reality suffer an *exceedingly* large number of individual accelerations by collisions with the molecules of the surrounding fluid. This second consideration would suggest, from analogy with the simple case of the problem of random flights [Eq. (108)], a formula of the *form* (320). The particular value of q (321) then follows from the first requirement.

Combining Eqs. (319) and (320) we obtain for the *transition probability* $\psi(\boldsymbol{u}; \delta\boldsymbol{u})$ for $\boldsymbol{u}$ to suffer an increment $\delta \boldsymbol{u}$ due to the Brownian forces only, the expression

$$\psi(\boldsymbol{u}; \delta\boldsymbol{u}) = \frac{1}{(4\pi q\Delta t)^{\frac{3}{2}}} \exp\left(-|\beta \boldsymbol{u}\Delta t + \delta \boldsymbol{u}|^2/4q\Delta t\right). \tag{322}$$

We shall now briefly re-examine the problem of continuous stochastic processes more generally from the point of view of the invariance of the *Maxwell-Boltzmann* distribution

$$W = \text{constant}\exp\{-[m|\boldsymbol{u}|^2 + 2m\mathfrak{B}(\boldsymbol{r})]/2kT\}; \quad \boldsymbol{K} = -\operatorname{grad}\mathfrak{B} \tag{323}$$

to processes governed by Eqs. (317) and (318) *only* i.e., without making the further assumptions included in Eqs. (319)–(322).

Assuming, as we have done hitherto, that the stochastic process we are considering is of the Markoff type we can write the integral equation [cf. Eq. (241)]

$$W(\boldsymbol{r}, \boldsymbol{u}, t+\Delta t) = \int\int W(\boldsymbol{r}-\Delta\boldsymbol{r}, \boldsymbol{u}-\Delta\boldsymbol{u}, t)\Psi(\boldsymbol{r}-\Delta\boldsymbol{r}, \boldsymbol{u}-\Delta\boldsymbol{u}; \Delta\boldsymbol{r}, \Delta\boldsymbol{u})d(\Delta\boldsymbol{r})d(\Delta\boldsymbol{u}). \tag{324}$$

According to Eqs. (318) we expect that [cf. Eq. (242)]

$$\Psi(\boldsymbol{r}, \boldsymbol{u}; \Delta\boldsymbol{r}, \Delta\boldsymbol{u}) = \psi(\boldsymbol{r}, \boldsymbol{u}; \Delta\boldsymbol{u})\delta(\Delta x - u_1\Delta t)\delta(\Delta y - u_2\Delta t)\delta(\Delta z - u_3\Delta t). \tag{325}$$

Equation (324) becomes

$$W(\boldsymbol{r}, \boldsymbol{u}, t+\Delta t) = \int W(\boldsymbol{r}-\boldsymbol{u}\Delta t, \boldsymbol{u}-\boldsymbol{K}\Delta t-\delta\boldsymbol{u}, t)\psi(\boldsymbol{r}-\boldsymbol{u}\Delta t, \boldsymbol{u}-\boldsymbol{K}\Delta t-\delta\boldsymbol{u}; \boldsymbol{K}\Delta t+\delta\boldsymbol{u})d(\delta\boldsymbol{u}), \tag{326}$$

where we have further substituted for $\Delta\boldsymbol{u}$ according to Eq. (317). Equation (326) can be written alternatively as

$$W(\boldsymbol{r}+\boldsymbol{u}\Delta t, \boldsymbol{u}+\boldsymbol{K}\Delta t, t+\Delta t) = \int W(\boldsymbol{r}, \boldsymbol{u}-\delta\boldsymbol{u}, t)\psi(\boldsymbol{r}, \boldsymbol{u}-\delta\boldsymbol{u}; \delta\boldsymbol{u})d(\delta\boldsymbol{u}). \tag{327}$$

Applying to this equation the same procedure as was adopted in the derivation of the Fokker-Planck and the generalized Liouville equations in §4, we readily find that [cf. Eq. (245)]

$$\left\{\frac{\partial W}{\partial t} + \boldsymbol{u}\cdot\operatorname{grad}_{\boldsymbol{r}} W + \boldsymbol{K}\cdot\operatorname{grad}_{\boldsymbol{u}} W\right\}\Delta t + O(\Delta t^2) = -\sum_i \frac{\partial}{\partial u_i}(W\langle\delta u_i\rangle_{\text{Av}}) + \frac{1}{2}\sum_i \frac{\partial^2}{\partial u_i^2}(W\langle\delta u_i^2\rangle_{\text{Av}})$$
$$+ \sum_{i<j}\frac{\partial^2}{\partial u_i\partial u_j}(W\langle\delta u_i\delta u_j\rangle_{\text{Av}}) + O(\langle\delta u_i\delta u_j\delta u_k\rangle_{\text{Av}}) \tag{328}$$

where $\langle\delta u_i\rangle_{\text{Av}}$ etc., denote the various moments of the transition probability $\psi(\boldsymbol{r}, \boldsymbol{u}; \delta\boldsymbol{u})$.

We shall now suppose that

$$\langle\delta u_i\rangle_{\text{Av}} = \mu_i\Delta t + O(\Delta t^2); \quad \langle\delta u_i^2\rangle_{\text{Av}} = \mu_{ii}\Delta t + O(\Delta t^2); \quad \langle\delta u_i\delta u_j\rangle_{\text{Av}} = \mu_{ij}\Delta t + O(\Delta t^2), \tag{329}$$

and that all averages of quantities like $\delta u_i\delta u_j\delta u_k$ are of order higher than one in Δt. With this understanding we shall obtain from Eq. (328), on passing to the limit $\Delta t = 0$ the result

$$\frac{\partial W}{\partial t} + \boldsymbol{u}\cdot\operatorname{grad}_{\boldsymbol{r}} W + \boldsymbol{K}\cdot\operatorname{grad}_{\boldsymbol{u}} W = -\sum_i \frac{\partial}{\partial u_i}(W\mu_i) + \frac{1}{2}\sum_i \frac{\partial^2}{\partial u_i^2}(W\mu_{ii}) + \sum_{i<j}\frac{\partial^2}{\partial u_i\partial u_j}(W\mu_{ij}). \tag{330}$$

We now require that the Maxwell-Boltzmann distribution (323) satisfy Eq. (330) identically. On substituting this distribution in Eq. (330) we find that the left-hand side of this equation vanishes and we are left with

$$-\sum_i \frac{\partial}{\partial u_i}[\exp(-m|\boldsymbol{u}|^2/2kT)\mu_i] + \frac{1}{2}\sum_i \frac{\partial^2}{\partial u_i^2}[\exp(-m|\boldsymbol{u}|^2/2kT)\mu_{ii}]$$
$$+ \sum_{i<j}\frac{\partial^2}{\partial u_i\partial u_j}[\exp(-m|\boldsymbol{u}|^2/2kT)\mu_{ij}] = 0. \tag{331}$$

Equation (331) is to be regarded as the general condition on the moments.

For the distribution (322)

$$\mu_i = -\beta u_i; \quad \mu_{ii} = 2q = 2\beta kT/m; \quad \mu_{ij} = 0. \tag{332}$$

Also, the third and higher moments of $\delta \mathbf{u}$ do not contain terms linear in Δt.

We readily verify that with the μ's given by (332) we satisfy Eq. (331). It is not, however, to be expected that (332) represents the most general solution for the μ's which will satisfy Eq. (331). It would clearly be a matter of considerable interest to investigate Eq. (331) (or the generalization of this equation to include terms involving μ_{ijk} etc.,) with a view to establishing the nature of the restrictions on the μ's implied by Eq. (331). Such an investigation might lead to the discovery of new classes of Markoff processes which will leave the Maxwell-Boltzmann distribution invariant but which will not be of the classical Brownian motion type. It is not proposed to undertake this investigation in this article. We may, however, draw special attention to the fact that according to Eqs. (331) and (332), β can very well depend on the spatial coordinates (though $q/\beta[=kT/m]$ must be a constant throughout the system). Thus, the generalized Liouville Eq. (249) and the Smoluchowski Eq. (312) are valid as they stand, also when $\beta = \beta(\mathbf{r})$.

CHAPTER III

PROBABILITY AFTER-EFFECTS: COLLOID STATISTICS; THE SECOND LAW OF THERMODYNAMICS. THE THEORY OF COAGULATION, SEDIMENTATION, AND THE ESCAPE OVER POTENTIAL BARRIERS

In this chapter we shall consider certain problems in the theory of Brownian motion which require the more explicit introduction than we had occasion hitherto, of the notion of *probability after-effects*. The fundamental ideas underlying this notion have already been described in the introductory section where we have also seen that colloid statistics (or, more generally, the phenomenon of density fluctuations in a medium of constant average density) provides a very direct illustration of the problem. The theory of this phenomenon which has been developed along very general lines by Smoluchowski has found beautiful confirmation in the experiments of Svedberg, Westgren, and others. This theory of Smoluchowski in addition to providing a striking application of the principles of Brownian motion has also important applications to the elucidation of the statistical nature of the second law of thermodynamics. In view, therefore, of the fundamental character of Smoluchowski's theory we shall give a somewhat detailed account of it in this chapter (§§1–3). (In the later sections of this chapter we consider further miscellaneous applications of the theory of Brownian motion which have bearings on problems considered in Chapter IV.)

1. The General Theory of Density Fluctuations for Intermittent Observations. The Mean Life and the Average Time of Recurrence of a State of Fluctuation

Consider a geometrically well-defined small element of volume v in a solution containing Brownian particles under conditions of diffusion equilibrium. (More generally, we may also consider v as an element in a very much larger volume containing a large number of particles in equilibrium.) Suppose now that we observe the number of particles contained in v systematically at constant intervals of time τ apart. Then the frequency $W(n)$ with which different numbers of particles will be observed in v will follow the *Poisson distribution* (see Appendix III),

$$W(n) = e^{-\nu}\nu^n/n!, \tag{333}$$

where ν denotes the average number of particles that will be contained in v:

$$\langle n \rangle_{\text{Av}} = \sum_{n=0}^{\infty} nW(n) = e^{-\nu}\nu \sum_{n=1}^{\infty} \frac{\nu^{n-1}}{(n-1)!} = \nu. \tag{334}$$

In other words, the number of particles that will be observed in v is subject to *fluctuations* and the different *states of fluctuations* (which, in this case, can be labelled by n) occur with definite frequencies.

According to Eq. (333) the mean square deviation δ^2 from the average value ν is given by

$$\delta^2 = \langle (n-\nu)^2 \rangle_{\text{Av}} = \langle n^2 \rangle_{\text{Av}} - \nu^2, \tag{335}$$

or, since

$$\left.\begin{aligned}\langle n^2\rangle_{\text{Av}}&=\sum_{n=0}^{\infty}n^2\frac{e^{-\nu}\nu^n}{n!}\\&=e^{-\nu}\nu\left\{\nu\sum_{n=2}^{\infty}\frac{\nu^{n-2}}{(n-2)!}+\sum_{n=1}^{\infty}\frac{\nu^{n-1}}{(n-1)!}\right\}\\&=\nu^2+\nu,\end{aligned}\right\}\quad(336)$$

we have

$$\delta^2=\nu. \quad (337)$$

It is seen that the frequency with which the different states of fluctuation n occur is independent of all physical parameters describing the particle (e.g., radius and density) and the surrounding fluid (e.g., viscosity). The situation is, however, completely changed when we consider the *speed* with which the different states of fluctuations follow each other in time. More specifically, consider the number of particles n and m contained in v at an interval of time τ apart. We expect that the number m observed on the second occasion will be correlated with the number n observed on the first occasion. This correlation should be such, that as $t\rightarrow 0$ the result of the second observation can be predicted with certainty as n, while as $t\rightarrow\infty$ we shall observe on the second occasion numbers which will increasingly be distributed according to the Poisson distribution (333). For finite intervals of time τ we can therefore ask for the *transition probability* $W(n;m)$ that m particles will be counted in v after a time τ from the instant when there was observed to be n particles in it.

In solving the problem stated toward the end of the preceding paragraph we shall make, following Smoluchowski, the two assumptions: (1) that the motions of the individual particles are not mutually influenced and are independent of each other and (2) that all positions in the element of volume considered have equal *a priori* probability. Under these circumstances we can expect to define *a probability P that a particle somewhere inside v will have emerged from it during the time τ*. The exact value of this *probability after-effect factor* P will depend on the precise circumstances of the problem including the geometry of the volume v. In §2 we shall obtain the explicit formula for P when the motions of the individual particles are governed by the laws of Brownian motion [Eq. (380)]; and similarly in §3 we shall obtain the formula for P for the case when the particles describe linear trajectories [Eq. (413)]. Meantime, we shall continue the discussion of the speed of fluctuations on the assumption that the factor P as defined can be unambiguously evaluated depending, however, on circumstances.

It is clear that the required transition probability $W(n;m)$ can be written down in an entirely elementary way if we know the probabilities with which particles enter and leave the element of volume. More precisely, let $A_i^{(n)}$ denote the probability that starting from an initial situation in which there are n particles inside v *some* i particles will have emerged from it during τ; this probability of emergence of a certain number of particles will clearly depend on the initial number of particles inside v. Similarly, let E_i denote the probability that i particles will have entered the element of volume v during τ. Since one of our principal assumptions is that the motions of the particles are not mutually influenced, the probability of entrance of a certain number of particles cannot depend on the number already contained in it. We shall now obtain explicit expressions for these two probabilities in terms of P.

The expression for $A_i^{(n)}$ can be written down at once when we recall that this must be equal to the product of the probability P^i that some particular group of i particles leaves v during τ, the probability $(1-P)^{n-i}$ that the remaining $(n-i)$ particles do not leave v during τ, and the number of distinct ways C_i^n of selecting i particles from the initial group of n. Accordingly,

$$A_i^{(n)}=C_i^nP^i(1-P)^{n-i}=\frac{n!}{i!(n-i)!}P^i(1-P)^{n-i}, \quad (338)$$

which is a Bernoulli distribution.

To obtain the expression for E_i we first remark that this must equal the probability that i particles *emerge* from the element of volume v on an *arbitrary* occasion; since, under *equilibrium conditions* the *a priori* probabilities for the entrance and emergence of particles must be equal. Remembering further, that E_i is independent of the number of particles initially contained in v, we clearly have

$$E_i = \langle A_i^{(n)} \rangle_{Av} = \sum_{n=i}^{\infty} W(n) A_i^{(n)}, \tag{339}$$

where $W(n)$ is the probability that v initially contained n particles; $W(n)$ accordingly is given by (333). Combining Eqs. (333), (338), and (339) we therefore have

$$\left.\begin{aligned} E_i &= \sum_{n=i}^{\infty} \frac{e^{-\nu}\nu^n}{n!} \frac{n!}{i!(n-i)!} P^i(1-P)^{n-i}, \\ &= \frac{e^{-\nu}(\nu P)^i}{i!} \sum_{n=i}^{\infty} \frac{\nu^{n-i}(1-P)^{n-i}}{(n-i)!}, \\ &= \frac{e^{-\nu}(\nu P)^i}{i!} e^{\nu(1-P)}. \end{aligned}\right\} \tag{340}$$

Thus,

$$E_i = e^{-\nu P}(\nu P)^i / i!, \tag{341}$$

in other words, a Poisson distribution with variance νP.

Using the formulae (338) and (341) for $A_i^{(n)}$ and E_i we can at once write down the expression for the transition-probability $W(n; n+k)$ that there is an increase in the number of particles from n to $n+k$. We clearly have

$$W(n; n+k) = \sum_{i=0}^{n} A_i^{(n)} E_{i+k}. \tag{342}$$

Similarly, for the transition probability $W(n; n-k)$ that there is a decrease in the number of particles from n to $n-k$ we have

$$W(n; n-k) = \sum_{i=k}^{n} A_i^{(n)} E_{i-k}, \quad (k \leqslant n). \tag{343}$$

From Eqs. (338), (341), (342), and (343) we therefore obtain

$$W(n; n+k) = e^{-\nu P} \sum_{i=0}^{n} C_i^n P^i (1-P)^{n-i} \times (\nu P)^{i+k}/(i+k)!, \tag{344}$$

and

$$W(n; n-k) = e^{-\nu P} \sum_{i=k}^{n} C_i^n P^i (1-P)^{n-i} \times (\nu P)^{i-k}/(i-k)! \tag{345}$$

The foregoing expressions for the transition probabilities are due to Smoluchowski.

The formulae (344) and (345) in spite of their apparent complexity have in reality very simple structures. To see this we first introduce the Bernoulli and the Poisson distributions

$$w_1^{(n)}(x) = C_x^n (1-P)^x P^{n-x} \quad (0 \leqslant x \leqslant n), \tag{346}$$

and

$$w_2(y) = e^{-\nu P}(\nu P)^y / y! \quad (0 \leqslant y < \infty). \tag{347}$$

$w_1^{(n)}(x)$ is the probability that *some* x particles *remain* in v after a time τ when initially there were n particles in it; similarly, $w_2(y)$ is the probability that y particles enter v in time τ. In terms of the distributions (346) and (347) we can rewrite Eqs. (344) and (345) as

$$W(n; n+k) = \sum_{i=0}^{n} w_1^{(n)}(n-i) w_2(i+k), \tag{348}$$

and

$$W(n; n-k) = \sum_{i=k}^{n} w_1^{(n)}(n-i) w_2(i-k), \tag{349}$$

or, writing m for $n+k$, respectively $n-k$, we see that both Eqs. (348) and (349) can be included in the single formula

$$W(n, m) = \sum_{x+y=m} w_1^{(n)}(x) w_2(y). \tag{350}$$

In other words, *the distribution* $W(n, m)$ for a fixed value of n *is the "sum" of the two distributions* (346) *and* (347). And, therefore, the mean and the mean square deviation for the distribution of m according to (350) is the sum of the means and the mean square deviations of the component distributions (346) and (347) (see Appendix IV). Since [cf. Eqs. (334) and (335) and Appendix I Eqs. (621) and (624)]

$$\langle x \rangle_{Av} = n(1-P); \quad \langle (x - \langle x \rangle_{Av})^2 \rangle_{Av} = nP(1-P), \tag{351}$$

and

$$\langle y \rangle_{Av} = \nu P; \quad \langle (y - \langle y \rangle_{Av})^2 \rangle_{Av} = \nu P, \tag{352}$$

we conclude that

$$\langle m \rangle_{Av} = n(1-P) + \nu P, \tag{353}$$

and

$$\langle (m - \langle m \rangle_{Av})^2 \rangle_{Av} = nP(1-P) + \nu P. \tag{354}$$

Let

$$\Delta_n = m - n. \tag{355}$$

Then, according to Eqs. (354) and (355)

$$\langle \Delta_n \rangle_{Av} = \langle m \rangle_{Av} - n = (\nu - n)P, \tag{356}$$

and

$$\left.\begin{aligned} \langle \Delta_n^2 \rangle_{Av} &= \langle (m - \langle m \rangle_{Av} + \langle m \rangle_{Av} - n)^2 \rangle_{Av} \\ &= \langle (m - \langle m \rangle_{Av})^2 \rangle_{Av} + (\langle m \rangle_{Av} - n)^2 \\ &= nP(1-P) + \nu P + (\nu - n)^2 P^2, \end{aligned}\right\} \tag{357}$$

or

$$\langle \Delta_n^2 \rangle_{Av} = P^2[(\nu - n)^2 - n] + (n + \nu)P. \tag{358}$$

It is seen that according to Eq. (356) the number of particles inside v changes, on the average, in the direction of making n approach its mean value, namely ν. In other words, the density fluctuations studied here in terms of a "microscopic" analysis of the stochastic motions of the individual particles are in complete agreement with the macroscopic theory of diffusion.

The quantities $\langle\Delta_n\rangle_{Av}$ and $\langle\Delta_n^2\rangle_{Av}$ represent the mean and the mean square of the differences that are to be expected in the numbers observed on two occasions at an interval of time τ apart when on the first occasion n particles were observed. If now, we further average $\langle\Delta_n\rangle_{Av}$ and $\langle\Delta_n^2\rangle_{Av}$ over all values of n with the weight function $W(n)$ we shall obtain the mean and the mean square of the differences in the numbers of particles observed on consecutive occasions in a long sequence of observations made at constant intervals τ apart. Thus [cf. Eq. (334)]

$$\langle\Delta\rangle_{Av}=\langle\langle\Delta_n\rangle_{Av}\rangle_{Av}=\langle\nu-n\rangle_{Av}P=0, \quad (359)$$

a result which is to be expected. On the other hand [cf. Eq. (337)]

$$\left.\begin{aligned}\langle\Delta^2\rangle_{Av}&=\langle\langle\Delta_n^2\rangle_{Av}\rangle_{Av}\\&=P^2[\langle(\nu-n)^2\rangle_{Av}-\langle n\rangle_{Av}]+\langle n+\nu\rangle_{Av}P\\&=P^2(\delta^2-\langle n\rangle_{Av})+(\langle n\rangle_{Av}+\nu)P,\end{aligned}\right\} \quad (360)$$

or

$$\langle\Delta^2\rangle_{Av}=2\nu P. \quad (361)$$

Equation (361) suggests a direct method for the experimental determination of the probability after-effect factor P from the simple evaluation of the mean square differences $\langle\Delta^2\rangle_{Av}$ from long sequences of observations of n (see §2 below). Further, according to Eq. (361)

$$\langle\Delta^2\rangle_{Av}=2\nu \quad \text{when} \quad P=1. \quad (362)$$

This result is in agreement with what we should expect, since, when $P=1$ there will be no correlation between the numbers that will be observed on two occasions at an interval τ apart; $\langle\Delta^2\rangle_{Av}$ then simply becomes the mean square of the differences between two numbers each of which (without correlation) is governed by the same Poisson distribution; and, therefore [cf. Eqs. (333) and (336)],

$$\begin{aligned}\langle\Delta^2\rangle_{Av}&=\langle(n-m)^2\rangle_{Av}=\langle n^2\rangle_{Av}+\langle m^2\rangle_{Av}-2\langle n\rangle_{Av}\langle m\rangle_{Av}\\&=2(\nu^2+\nu)-2\nu^2=2\nu, \quad (P=1).\end{aligned} \quad (363)$$

We shall now show how we can define the *mean life* and *the average time of recurrence* for a given state of fluctuation in terms of the transition probability $W(n;n)$:

$$W(n;n)=e^{-\nu P}\sum_{i=0}^{n}C_i^nP^i(1-P)^{n-i}(\nu P)^i/i!, \quad (364)$$

which gives the probability that n will be observed on two consecutive occasions. Accordingly, the probability $\phi_n(k\tau)$ that the same number n will be observed on $(k-1)$ consecutive occasions (at constant intervals τ apart) and that on the kth occasion some number different from n will be observed is given by

$$\phi_n(k\tau)=W^{k-1}(n;n)[1-W(n;n)]. \quad (365)$$

On the other hand, in terms of $\phi_n(k\tau)$ we can give a natural definition to the mean life to the state of fluctuation n by the equation

$$T_n=\sum_{k=1}^{\infty}k\tau\phi_n(k\tau). \quad (366)$$

Combining Eqs. (365) and (366) we obtain

$$T_n=\tau[1-W(n;n)]\sum_{k=1}^{\infty}kW^{k-1}(n;n). \quad (367)$$

The infinite series in Eq. (367) is readily evaluated and we find

$$T_n=\frac{\tau}{1-W(n;n)}. \quad (368)$$

In an analogous manner we can define the time of recurrence of the state n by the equation

$$\Theta_n=\sum_{k=1}^{\infty}k\tau\psi_n(k\tau), \quad (369)$$

where $\psi_n(k\tau)$ denotes the probability that starting from *an arbitrary state which is not n* we shall observe on $k-1$ successive occasions states which are not n and on the kth occasion observe the state n. If

$$W(Nn;Nn) \quad (370)$$

denotes the probability that from an arbitrary state $\neq n$ we shall have a transition to a state which is also $\neq n$, then clearly

$$\psi_n(k\tau)=W^{k-1}(Nn;Nn)[1-W(Nn;Nn)]. \quad (371)$$

Substituting the foregoing expression for $\psi_n(k\tau)$ in Eq. (369) we obtain [cf. Eqs. (365) and (368)]

$$\Theta_n=\frac{\tau}{1-W(Nn;Nn)}. \tag{372}$$

We shall now obtain a formula for $W(Nn;Nn)$. First of all it is clear that

$$1-W(Nn;Nn)=W(Nn;n), \tag{373}$$

where $W(Nn;n)$ is the probability that from an arbitrary state $\neq n$ we shall have a transition to the state n. Now, under equilibrium conditions, the number of transitions from states $\neq n$ to the state n must equal the number of transisitions from the state n to states $\neq n$; accordingly

$$[1-W(n)]W(Nn;n)=W(n)[1-W(n;n)], \tag{374}$$

where $W(n)$ is given by Eq. (333). Hence,

$$W(Nn;n)=W(n)\frac{1-W(n;n)}{1-W(n)}. \tag{375}$$

Combining Eqs. (372), (373), and (375) we obtain

$$\Theta_n=\frac{\tau}{1-W(n;n)}\frac{1-W(n)}{W(n)}. \tag{376}$$

Finally, we may note that between T_n and Θ_n we have the relation

$$\Theta_n=T_n\frac{1-W(n)}{W(n)}. \tag{377}$$

In the next section we shall give a brief account of the experiments of Svedberg and Westgren on colloid statistics which have provided complete confirmation of Smoluchowski's theory of density fluctuations which we have developed in this section. Also, the formulae for T_n and Θ_n which we have derived have important applications to the elucidation of the second law of thermodynamics to which we shall return in §4.

2. Experimental Verification of Smoluchowski's Theory: Colloid Statistics

In the experiments of Svedberg, Westgren, and others on colloid statistics observations are

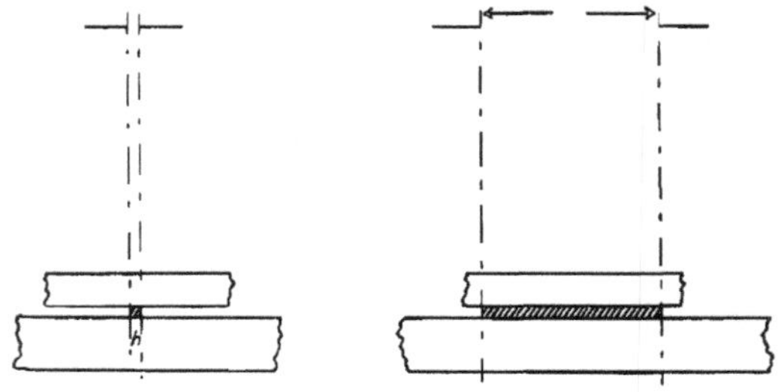

FIG. 3.

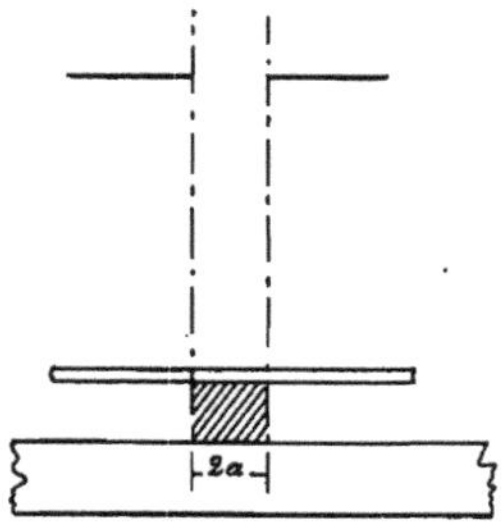

FIG. 4.

made by means of an ultramicroscope on the numbers of particles in a well-defined element of volume in a colloidal solution. These observations, made systematically at constant intervals τ apart, are secured either by the use of intermittent illumination (Svedberg) or by counting on the ticks of a metronome (Westgren). The volumes in which the counts are made are defined either optically by illuminating only plane parallel layers several microns in thickness (Svedberg) or mechanically by having the solution under observation sealed between the objective of the microscope and a glass plate and observing with the help of a cardioid condenser (Westgren). The dimensions of the element of volume at right angles to the line of sight are defined directly by limiting the field of observation (see Figs. 3 and 4).

The colloidal particles describe Brownian motion and since the intervals of time we are normally interested in are never less than a few hundredths of a second we can suppose that the motions of the particles are governed by the diffusion equation [cf. Eqs. (133), (304), and (306)]

$$\partial w/\partial t=D\nabla^2 w;\quad D=q/\beta^2=kT/m\beta=kT/6\pi a\eta. \tag{378}$$

For, according to our discussion in Chapter II, §§2 and 4 the validity of the diffusion equation requires that we only ignore what happens in time intervals of order β^{-1} and for colloidal gold particles of radius $a=50\mu\mu$ this time of relaxation is of the order of 10^{-9}–10^{-10} second.

From Eq. (378) we conclude that the probability of occurrence of a particle at $\mathbf{r}_2$ at time t when it was at $\mathbf{r}_1$ at time $t=0$ is given by [cf. Eq. (172)]

$$\frac{1}{(4\pi Dt)^{\frac{3}{2}}}\exp\left(-|\mathbf{r}_2-\mathbf{r}_1|^2/4Dt\right). \tag{379}$$

On this basis we can readily write down a general formula for the probability after-effect factor P introduced in §1. For, by definition, P denotes the probability that a particle somewhere inside the given element of volume v (with uniform probability) at time $t=0$ will find itself outside of it at time $t=\tau$. Accordingly

$$P=\frac{1}{(4\pi D\tau)^{\frac{3}{2}}v}\int\int\exp\left(-|\mathbf{r}_1-\mathbf{r}_2|^2/4D\tau\right)d\mathbf{r}_1d\mathbf{r}_2, \tag{380}$$

where the integration over $\mathbf{r}_1$ is extended over all points in the interior of v while that over $\mathbf{r}_2$ is extended over all points exterior to v. Alternatively, we can also write

$$1-P=\frac{1}{(4\pi D\tau)^{\frac{3}{2}}v}\int_{\mathbf{r}_1\epsilon v}\int_{\mathbf{r}_2\epsilon v}\times\exp\left(-|\mathbf{r}_1-\mathbf{r}_2|^2/4D\tau\right)d\mathbf{r}_1d\mathbf{r}_2, \tag{381}$$

where, now, the integrations over *both* $\mathbf{r}_1$ and $\mathbf{r}_2$ are extended over all points inside v (indicated by the symbols $\mathbf{r}_1\epsilon v$ and $\mathbf{r}_2\epsilon v$).

We thus see that for any geometrically well-defined element of volume in a colloidal solution we can always evaluate, in principle, the probability after-effect factor P in terms of the physical parameters of the problem, namely, the geometry of the volume v, the radius a of the colloidal particles, and the coefficient of viscosity η of the surrounding liquid. On the other hand, this factor P can also be determined empirically from a direct evaluation of the mean square of the differences in the numbers of particles observed on consecutive occasions in a long sequence of observations made at constant intervals τ apart and using the formula [Eq. (361)]

$$\langle\Delta^2\rangle_{\text{Av}}=2\nu P, \tag{382}$$

where ν is the average of all the numbers observed. A comparison of the predictions of the theory with the data of colloid statistics therefore becomes possible. Once P has been determined [either theoretically according to Eq. (381) or empirically from Eq. (382)] we can predict the frequency of occurrence, $H(n, m)$, of the *pair* (n, m) in the observed sequence of numbers. For, clearly;

$$H(n, m)=W(n)W(n;m), \tag{383}$$

where $W(n)$ is the frequency of occurrence of n according to Eq. (333) and $W(n;m)$ is the transition probability from the state n to the state m according to Smoluchowski's formulae (344) and (345). Again a comparison between the predictions of the theory with the results of observations becomes possible.

Comparisons of the kind indicated in the preceding paragraph were first made by Smoluchowski himself who used for this purpose the data provided by Svedberg's experiments. However, later experiments by Westgren carried out with the expressed intention of verifying Smoluchowski's theory provide a more stringent comparison between the predictions of the theory and the results of observations. We shall therefore limit ourselves to describing the results of Westgren's experiments only.

Westgren conducted two series of experiments with the arrangements shown in Figs. 3 and 4. In the first of the two arrangements (Fig. 3) the particles under observation are confined to a long rectangular parallelepiped (see the shaded portions in Fig. 3). Under the conditions of this arrangement it is clear that the variation in the number of particles observed is predominantly due to diffusion at right angles to the lengthwise edge. Consequently, the formula for P appropriate to this arrangement is [cf. Eq. (381)]

$$1-P=\frac{1}{h(4\pi D\tau)^{\frac{1}{2}}}\int_0^h\int_0^h\times\exp\left[-(x_1-x_2)^2/4D\tau\right]dx_1dx_2, \tag{384}$$

where h denotes the width of the element of

volume under observation (see Fig. 3). Introducing $2(D\tau)^{\frac{1}{2}}$ as the unit of length, Eq. (384) becomes

$$1-P=\frac{1}{\alpha\pi^{\frac{1}{2}}}\int_0^\alpha\int_0^\alpha \exp\left[-(\xi_1-\xi_2)^2\right]d\xi_1 d\xi_2, \quad (385)$$

where we have written

$$\alpha=h/2(D\tau)^{\frac{1}{2}}. \quad (386)$$

We readily verify that Eq. (385) is equivalent to

$$1-P=\frac{2}{\alpha\pi^{\frac{1}{2}}}\int_0^\alpha d\xi_1\int_0^{\xi_1} d\eta \exp(-\eta^2), \quad (387)$$

or, after an integration by parts we find

$$P=1-\frac{2}{\pi^{\frac{1}{2}}}\int_0^\alpha \exp(-\xi^2)d\xi + \frac{1}{\alpha\pi^{\frac{1}{2}}}\left[1-\exp(-\alpha^2)\right]. \quad (388)$$

For the second of Westgren's arrangements (Fig. 4) the element under observation is a cylindrical volume and the variations in the numbers observed are in this case due to the diffusion of particles in all directions at right angles to the line of sight. Accordingly we have

$$P=\frac{4}{\alpha^2\pi}\int_\alpha^\infty d\xi_1\xi_1\int_0^\alpha d\xi_2\xi_2\int_0^\pi \times\exp(-\xi_1^2-\xi_2^2+2\xi_1\xi_2\cos\vartheta)d\vartheta, \quad (389)$$

where

$$\alpha=r_0/2(D\tau)^{\frac{1}{2}}, \quad (390)$$

r_0 denoting the radius of the cylindrical element under observation. The integrals in (389) can be evaluated in terms of Bessel functions with imaginary arguments and we find

$$P=e^{-2\sqrt{\alpha}}\left[I_0(2\sqrt{\alpha})+I_1(2\sqrt{\alpha})\right]. \quad (391)^6$$

Westgren has made several series of counts with both of his experimental arrangements. We give below a sample extract from one of his sequences:

2 1 1 1 1 1 0 2 2 1 1 1 2 3 2 3 0 0 0 0 0 0 1 1 0 0 1 0 1 1 1 2 2 3 3 4 5
3 4 2 2 1 2 1 3 2 0 2 2 1 0 2 2 2 1 2 3 2 2 2 3 2 2 2 2 2 2 1 3 3 4 2 2 (392)

The foregoing counts were obtained with the first of the two experimental arrangements described with the following values for the various physical parameters:

$$h=6.56\mu;\quad D=3.95\times10^{-8};\quad \tau=1.39\text{ sec.};\quad a=49.5\mu\mu;\quad T=290.0°\text{K};\quad \nu=1.428. \quad (393)$$

First of all, it is of interest to see how well the Poisson distribution (333) represents the observed frequencies of occurrence of the different values of n. Table III shows this comparison for the sequence of which (392) is an extract. It is seen that the representation is satisfactory. Also, the observed mean square deviation for this sequence is 1.35 while the value theoretically predicted is ν which is 1.43; again the agreement is satisfactory.

Turning next to questions relating to probability after-effects we may first note that each of the observed sequences can be used for several comparisons. For, by suitably selecting from a given sequence of sufficient length we can derive others with intervals between consecutive observations which are integral multiples of that characterizing the original sequence. Thus, by considering only the alternate numbers we obtain a new sequence in which the interval τ between two observations is twice that in the original sequence.

As we have already remarked, for any given sequence, we can compute theoretical values of P in terms of the physical parameters of the problem according to Eq. (388) or (391) depending on the experimental arrangement used. For the same sequences, we can also, using Eq. (382), derive values of P from the observed counts

TABLE III. The Poisson distribution for $W(n)$. $\nu=1.428$.

$n=$	0	1	2	3	4	5	6	7
$W(n)_{\text{obs}}$	381	568	357	175	67	28	5	2
$W(n)_{\text{calc}}$	380	542	384	184	66	19	5	2

[6] The functions $e^{-x}I_{0,1}(x)$ are tabulated in Watson's *Bessel functions* (Cambridge, 1922), pp. 698–713.

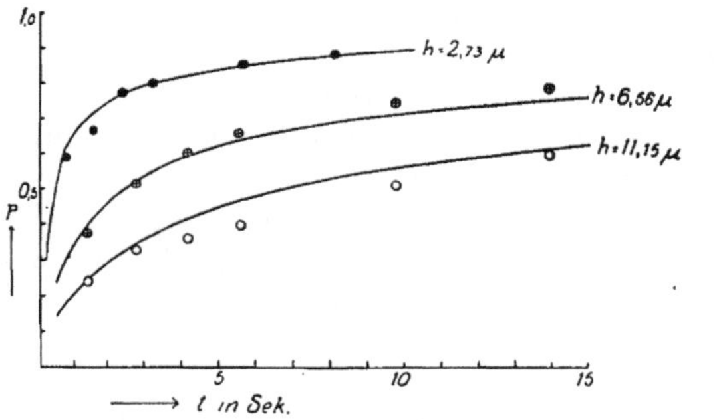

FIG. 5.

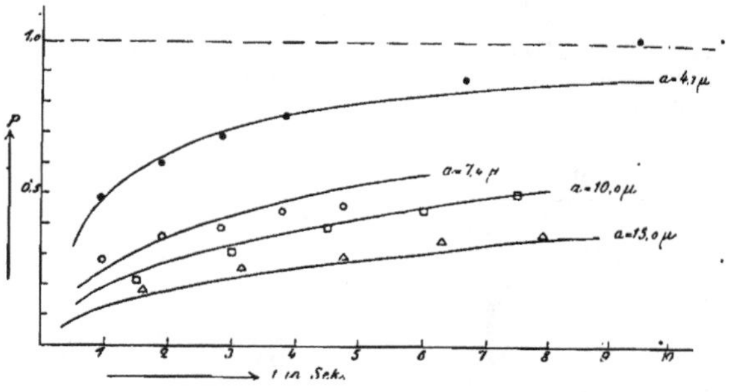

FIG. 6.

from the respective values of the mean square differences $\langle\Delta^2\rangle_{Av}$. In Tables IV and V we have made, following Westgren, the comparison between the values P derived in this manner for two typical cases. The agreement is satisfactory. The confirmation of the theory is shown in a particularly striking manner in Figs. 5 and 6 where a comparison is made between the observed and the theoretical values of P in its dependence on τ for different values of h (or r_0).

It is now seen that an analysis of the data of colloid statistics actually provides us with a means of determining the Avogadro's constant N. For, from the mean square difference $\langle\Delta^2\rangle_{Av}$ and the mean value of n (namely ν) we can determine P. On the other hand, according to Eqs. (380) and (381) P is determined by the geometry of the volume v *only*, *if* the unit of length is chosen to be $2(D\tau)^{\frac{1}{2}}$. Hence, from the empirically determined value of P we can deduce a value for this unit of length. In other words, a determination of the diffusion coefficient D is possible. But [cf. Eq. (378)]

$$D = kT/6\pi a\eta = (R/N)(T/6\pi a\eta), \qquad (394)$$

where R is the gas constant and N the Avogadro's number. Thus N can be determined. With the second of his two arrangements Westgren has used this method to determine N. As a mean of 50 determinations he finds $N = 6.09 \times 10^{23}$ with a probable error of 5 percent; this is in very satisfactory agreement with other independent determinations.

Turning next to the frequency of occurrence $H(n, m)$ of the pair of numbers (n, m) in a given sequence, we can predict this quantity according to Eq. (383); these predicted values can again be compared with those deduced directly from the counts. Such a comparison has also been made by Westgren whose results we give in Table VI.

Finally, we shall consider the experimental basis for the formulae (368) and (376) for the mean life and the average time of recurrence of a state of fluctuation. Using the counts of Svedberg, Smoluchowski has made a comparison between the values of T_n and Θ_n derived empirically from these counts and those predicted by Eqs. (368) and (376). The results of this comparison are shown in Table VII.

The long average times of recurrence for the states of large n are to be particularly noted (see §4 below). These long times are, however, a direct consequence of the "improbable"

TABLE IV. Comparison of the probability after-effect factor P derived from Eq. (388) and the experimental arrangement of Fig. 3 (Westgren). $h = 6.56\mu$; $a = 49.5\mu\mu$; $T = 290.0°K$; $D = 3.95 \times 10^{-8}$; $\nu = 1.428$.

τ(sec.)	$\langle\Delta^2\rangle_{Av}$	P_{obs}	P_{calc}
1.39	1.068	0.374	0.394
2.78	1.452	0.513	0.517
4.17	1.699	0.600	0.587
5.56	1.859	0.656	0.634
9.73	2.125	0.744	0.713
13.90	2.265	0.793	0.760

TABLE V. Comparison of the probability after-effect factor P derived from Eq. (391) and the experimental arrangement of Fig. 4 (Westgren). $r_0 = 10.0\mu$; $a = 63.5\mu\mu$; $T = 290.1°K$; $D = 3.024 \times 10^{-8}$; $\nu = 1.933$.

τ(sec.)	$\langle\Delta^2\rangle_{Av}$	P_{obs}	P_{calc}
1.50	0.836	0.217	0.238
3.00	1.200	0.310	0.332
4.50	1.512	0.391	0.401
6.00	1.718	0.444	0.456
7.50	1.939	0.502	0.503

TABLE VI. The observed and the theoretical frequencies of occurrence of the pairs (n, m) in a given sequence ($\nu=1.428$; $P=0.374$). [In each case the top figure gives the observed number while the bottom figure (italicized), the number to be expected on the basis of Eq. (383).]

n	$m=0$	1	2	3	4	5	6
0	210	126	35	7	0	1	—
	221	*119*	*32*	*6*	*1*	—	—
1	134	281	117	29	1	1	—
	119	*262*	*122*	*31*	*5*	*1*	—
2	27	138	108	63	16	3	—
	32	*122*	*149*	*63*	*15*	*3*	—
3	10	20	76	38	24	6	0
	6	*31*	*63*	*56*	*22*	*5*	*1*
4	2	2	14	22	13	11	3
	1	*5*	*15*	*22*	*15*	*6*	*2*
5	—	0	2	10	10	1	3
	—	*1*	*3*	*5*	*6*	*3*	*1*

nature of these states. For, according to Eq. (376)

$$\Theta_n \sim (\tau/W(n)) = \tau(e^{\nu} n!/\nu^n) \quad (n \gg \nu). \quad (395)$$

which increases extremely rapidly for large values of n. For example, the number 7 was recorded only once in Svedberg's entire sequence of 518 counts; but the average time of recurrence for this state is 1105τ. Again, the number 17 (for instance) was never observed by Svedberg; and this is also understandable in view of the average time of recurrence for this state which is $\Theta_{17} \sim 10^{13}\tau$!

In concluding this discussion of the experimental verification of Smoluchowski's theory, we may remark on the inner relationships that have been disclosed to exist between the phenomena of Brownian motion, diffusion, and fluctuations in molecular concentration. But what is perhaps of even greater significance is that we have here the first example of a case in which it has been possible to follow in all its details, both theoretically and experimentally, the transition between the macroscopically irreversible nature of diffusion and the microscopically reversible nature of molecular fluctuations. (These matters are further touched upon in §§3 and 4 below.)

3. Probability After-Effects for Continuous Observation

The theory of density fluctuations as developed in §1 is valid whenever the physical circumstances of the problem will permit us to introduce the probability after-effect factor P. It will be recalled that this factor $P(\tau)$ is defined as the probability that a particle, initially, somewhere inside a given element of volume will emerge from it before the elapse of a time τ. And, as we have seen in §1, we can express all the significant facts related to the phenomenon of the speed of fluctuations in terms of this single factor $P(\tau)$. But the theory as developed in §1 applies only when τ is finite, i.e., for the case of intermittent observations. We shall now show how this theory can be generalized to include the case of continuous observations.

First of all, it is clear that we should expect

$$P(\tau) \to 0 \quad \text{as} \quad \tau \to 0. \quad (396)$$

Hence, according to Eq. (364),

$$W(n;n) = e^{-\nu P}(1-P)^n + O(P^2) \quad (\tau\to 0;\ P\to 0), \quad (397)$$

or

$$W(n;n) = 1-(n+\nu)P(\tau)+O(P^2) \quad (\tau\to 0;\ P\to 0). \quad (398)$$

From this expression for $W(n;n)$ we can derive a formula for the probability $\phi_n(t)\Delta t$ that the state n will continue to be under observation for a time t and that during t and $t+\Delta t$ there will occur a transition to a state different from n. For this purpose, we divide the interval $(0, t)$ into a very large number of subintervals of duration Δt. Then, from the definition of $\phi_n(t)\Delta t$ it follows that

$$\phi_n(t)\Delta t = [W(n;n)]^{t/\Delta t}[1-W(n;n)], \quad (399)$$

or, using Eq. (398),

$$\phi_n(t)\Delta t = [1-(n+\nu)P(\Delta t)+O(P^2)]^{t/\Delta t} \times (n+\nu)P(\Delta t). \quad (400)$$

This last equation suggests that to obtain consistent results it would be necessary that

$$P(\Delta t) = O(\Delta t) \quad (\Delta t \to 0). \quad (401)$$

On general physical grounds, we may expect that this would in fact be the case. But it should not be concluded that Eq. (401) will be valid for *any* arbitrary idealization of the physical problem. For example, it is *not* true that $P(\Delta t)$ is $O(\Delta t)$ for the case of Brownian motions *idealized* as a problem in pure diffusion as we have done in §2. For, according to Eq. (379)

$$\langle |\Delta \mathbf{r}|^2 \rangle_{Av} = 6D\Delta t; \tag{402}$$

and hence, for P defined as in Eq. (380)

$$P = O[(\Delta t)^{\frac{1}{2}}] \quad (\Delta t \to 0), \tag{403}$$

contrary to Eq. (401). However, the reason for this disagreement is that the reduction of the problem of Brownian motions to one in diffusion can be achieved only when the intervals of time we are interested in are long compared to the time of relaxation β^{-1}. When this ceases to be the case, as in the present context, Eq. (379) is no longer true and we should strictly use the general distribution derived in Chapter II, §2 [see Eq. (171)]. And, according to Eq. (175)

$$\langle |\Delta \mathbf{r}|^2 \rangle_{Av} = |\mathbf{u}_0|^2 (\Delta t)^2, \quad (\Delta t \to 0). \tag{404}$$

On the basis of Eq. (404) we shall naturally be led to a formula for P consistent with (401) [see Eq. (413) below]. We shall therefore assume that

$$P(\Delta t) = P_0 \Delta t + O(\Delta t^2) \quad (\Delta t \to 0), \tag{405}$$

where P_0 is a constant.

Combining Eqs. (400) and (405) we have

$$\phi_n(t)\Delta t = [1 - (n+\nu)P_0\Delta t + O(\Delta t^2)]^{t/\Delta t} \times (n+\nu)P_0\Delta t, \tag{406}$$

or, passing to the limit $\Delta t = 0$ we obtain

$$\phi_n(t)dt = \exp[-(n+\nu)P_0 t](n+\nu)P_0 dt. \tag{407}$$

Equation (407) expresses *a law of decay of a state of fluctuation* quite analogous to the law of decay of radioactive substances.

According to Eq. (407), the mean life, T_n, of the state n for continuous observation can be defined by

$$T_n = \int_0^\infty t\phi_n(t)dt; \tag{408}$$

in other words

$$T_n = 1/(n+\nu)P_0. \tag{409}$$

Equation (409) is our present analogue of the formula (368) valid for intermittent observations.

Again, as in §1, we can also define the average time of recurrence of a state of fluctuation for continuous observation. This can be done by introducing the probability $W(Nn; Nn)$ and proceeding exactly as in the discussion of T_n. However, without going into details, it is evident that the relation (377) between T_n and Θ_n must continue to be valid, also for the case of continuous observation. Hence

$$\Theta_n = \frac{1}{(n+\nu)P_0} \frac{1 - W(n)}{W(n)}. \tag{410}$$

We shall now derive for the case of Brownian motions, an explicit formula for P_0 which we formally introduced in Eq. (405). As we have already remarked, when dealing with continuous observation, the idealization of the phenomenon of Brownian motion as pure diffusion is not tenable. Instead, we should base our discussion on the exact distribution function $W(\mathbf{r}, t; \mathbf{r}_0, \mathbf{u}_0)$ given by Eq. (171) and which is valid also for times of the order of the time of relaxation β^{-1}. However, since we are only interested in $P(\Delta t)$ for $\Delta t \to 0$ it would clearly be sufficient to consider the limiting form of the exact distribution $W(\mathbf{r}, \mathbf{u}, t; \mathbf{r}_0, \mathbf{u}_0)$ as $t \to 0$. On the other hand according to Eqs. (170)–(175) it follows that as $t \to 0$ we can regard the particles as describing linear trajectories with a Maxwellian distribution of the velocities. Hence, in our present context, $P(\Delta t)$ represents the probability that a particle initially inside a given element of volume v (with uniform probability) and with a velocity distribution governed by Maxwell's law will emerge from v before a time Δt. It is clear that formally, this is the same as the number of molecules striking the inner surface of the element of volume considered in a time Δt when the molecular concentration is $1/v$.

TABLE VII. The mean life T_n and the average time of recurrence Θ_n ($P = 0.726$; $\nu = 1.55$). (T_n and Θ_n are expressed in units of τ.)

n	T_n(obs.)	T_n(calc.)	Θ_n(obs.)	Θ_n(calc.)
0	1.67	1.47	6.08	5.54
1	1.50	1.55	3.13	3.16
2	1.37	1.38	4.11	4.05
3	1.25	1.23	7.85	8.07
4	1.23	1.12	18.6	20.9

Now, according to calculations familiar in the kinetic theory of gases, the number of molecules with velocities between $|\mathbf{u}|$ and $|\mathbf{u}|+d|\mathbf{u}|$ which strike unit area of any solid surface per unit time and in a direction with a solid angle $d\Omega$ at an angle ϑ with the normal to the surface is given by

$$N(m/2\pi kT)^{\frac{3}{2}} \exp(-m|\mathbf{u}|^2/2kT) \times|\mathbf{u}|^3 \cos\vartheta d\Omega d|\mathbf{u}|, \quad (411)$$

where N denotes the molecular concentration. Hence,

$$P(\Delta t)=\Delta t \frac{\sigma}{v}\left(\frac{m}{2\pi kT}\right)^{\frac{3}{2}} \int_0^\infty \int_0^\pi \exp(-m|\mathbf{u}|^2/2kT) \times|\mathbf{u}|^3 \cos\vartheta d\Omega d|\mathbf{u}|, \quad (412)$$

where σ is the total surface area of the element of volume v. On evaluating the integrals in Eq. (412) we find that

$$P(\Delta t)=(\sigma/v)(kT/2\pi m)^{\frac{1}{2}}\Delta t. \quad (413)$$

Comparing this with Eq. (405) we conclude that for the case under consideration

$$P_0=(\sigma/v)(kT/2\pi m)^{\frac{1}{2}}. \quad (414)$$

The formulae (409) and (410) for the mean life and the average time of recurrence now take the forms

$$T_n=(v/\sigma(n+\nu))(2\pi m/kT)^{\frac{1}{2}}, \quad (415)$$

and

$$\Theta_n=(v/\sigma(n+\nu))(2\pi m/kT)^{\frac{1}{2}} \times([1-W(n)]/W(n)). \quad (416)$$

The case of greatest interest arises when the average number of particles, ν, contained in v is a very large number and the values of n considered are relatively close to ν. Then, the Poisson distribution $W(n)$ simplifies to (see Appendix III)

$$W(n)=[1/(2\pi\nu)^{\frac{1}{2}}]\exp[-(n-\nu)^2/2\nu]. \quad (417)$$

On this approximation, Eq. (416) becomes

$$\Theta_n \simeq \pi \frac{v}{\sigma}\left(\frac{m}{\nu kT}\right)^{\frac{1}{2}} \exp[(n-\nu)^2/2\nu]. \quad (418)$$

As an illustration of Eq. (418) we shall consider, following Smoluchowski, the average time of recurrence of a state of fluctuation in which the molecular concentration of oxygen in a sphere of air of radius a will differ from the average value by 1 percent. Table VIII gives Θ_n for different values of a.

TABLE VIII. The average time of recurrence of a state of fluctuation in which the molecular concentration in a sphere of air of radius a will differ from the average value by 1 percent. $T=300°K$; $\nu=3\times10^{19}\times(4\pi a^3/3)$.

a(cm)	1	5×10^{-5}	3×10^{-5}	2.5×10^{-5}	1×10^{-5}
Θ(sec.)	$10^{10^{14}}$	10^{68}	10^{6}	1	10^{-11}

It is seen from Table VIII that under normal conditions, for volumes which are on the edge of visual perception even appreciable fluctuations in the molecular concentrations require such colossal average times of recurrence, that for all practical purposes the phenomenon of diffusion can be regarded as an irreversible process. On the other hand, for volumes which are just on the limit of microscopic vision, fluctuations in concentrations occur to such an extent and with such frequency that there can no longer be any question of irreversibility: under such conditions the notion of diffusion very largely loses its common meaning. For example, it would scarcely occur to one to illustrate the phenomenon of diffusion by the experiments of Svedberg and Westgren on colloid statistics though it is in fact true that *on the average* the results are in perfect accord with the principles of macroscopic diffusion [as is illustrated, for example, by Eq. (356) for $\langle\Delta_n\rangle_{Av}$]. We shall return to these questions in the following section.

4. On the Reversibility of Thermodynamically Irreversible Processes, the Recurrence of Improbable States, and the Limits of Validity of the Second Law of Thermodynamics

If we formulate the second law of thermodynamics in any of its conventional forms, as, for example, that "heat cannot of itself be transferred from a colder to a hotter body" or, that "arbitrarily near to any given state there exist states which are inaccessible to the initial state by adiabatic processes" (Caratheodory), or that "the entropy of a closed system must never decrease," we, at once, get into contradiction with the kinetic molecular theory which demands the essential reversibility of all processes. Consequently, from the side of "dogmatic" thermodynamics two principal objections have been raised in the form of paradoxes and which are held to vitiate the entire outlook of the kinetic theory and statistical mechanics. We first state the two paradoxes.

(*i*) *Loschmidt's Reversibility Paradox*

Loschmidt first drew attention to the fact that in view of the essential symmetry of the laws of mechanics to the past and the future, all molecular processes must be reversible from the point of view of statistical mechanics. This is in apparent contradiction with the point of view held in thermodynamics that certain processes are irreversible.

(*ii*) *Zermelo's Recurrence Paradox*

There is a theorem in dynamics due to Poincaré which states that *in a system of material particles under the influence of forces which depend only on the spatial coordinates, a given initial state*[7] *must, in general, recur, not exactly, but to any desired degree of accuracy, infinitely often, provided the system always remains in the finite part of the phase space.* (For a proof of this theorem see Appendix V.) In other words, the trajectory described by the representative point in the phase space has a "quasi-periodic" character in the sense that after a finite interval of time (which can be specified) the system will return to the initial state to any desired degree of accuracy. Basing on this theorem of Poincaré, Zermelo has argued that the notion of irreversibility fundamental to macroscopic thermodynamics is incompatible with the standpoint of the kinetic theory.

As is well known, Boltzmann has tried to resolve these paradoxes of Loschmidt and Zermelo by probability considerations of a general nature. Thus, on the strength of certain rough estimates (see Appendix VI), Boltzmann concludes that the period of one of Poincaré's cycles is so enormously long, even for a cubic centimeter of gas, that the recurrence of an initially improbable state (i.e., the reversal to a state of lower entropy) while not strictly impossible, is yet so highly improbable that during the times normally available for observation, the chance of witnessing a thermodynamically irreversible process is *extremely* small.

Though Boltzmann's arguments and conclusions are fundamentally sound there are certain unsatisfactory features in basing on the period of a Poincaré cycle. For one thing, the period of such a cycle depends on how *nearly* we (arbitrarily) require the initial state to recur. Again, Poincaré's theorem refers to the return of the representative point in the $6N$-dimensional phase space (N denoting the number of particles in the system). Actually, in practice, we should treat two states of a gas as macroscopically distinct only if the numbers of molecules (considered indistinguishable) in the various limits of positions and velocities are different. Then, during a Poincaré cycle, the different macroscopically distinguishable states of the system will approximately recur a great many times. These recurrences of the different macroscopically distinct states, during a given Poincaré cycle, will be distributed very unequally among the states: thus, most of the recurrences will occur for the states of the system which are very close to what would be described as the thermodynamically "*normal state.*" Moreover, it can also happen that during such a cycle, states deviating by arbitrarily large amounts from the normal state are assumed by the system. In other words, during a Poincaré cycle we shall pass through many improbable states and indeed with equal frequency both in the directions of increasing and decreasing entropy.

Thus, while we may accept Boltzmann's point of view as fundamentally correct, it would clearly add to our understanding of the whole problem if we can explicitly demonstrate in a given instance how in spite of the essential reversibility of all molecular phenomena, we nevertheless get the impression of irreversibility.

Now, as we have already remarked in the preceding sections, Smoluchowski's theory of fluctuations in molecular concentrations allows us to bridge the gap between the regions of the

[7] This is defined by the positions and the velocities of all the particles, i.e., by the representative point in the phase space.

macroscopically irreversible diffusion and the microscopically reversible fluctuations. Consequently, a further discussion of this problem will enable us to follow explicitly how in this particular instance the Loschmidt and the Zermelo paradoxes resolve themselves.

(*a*) *The resolution of Loschmidt's paradox.*—Using Eqs. (333), (344), and (345) we readily verify that

$$H(n, n+k) = W(n)W(n; n+k) = W(n+k)W(n+k; n) = H(n+k, n). \quad (419)$$

The quantity on the left-hand side in the foregoing equation represents the frequency of occurrence of the numbers n and $n+k$ on two successive occasions in a long sequence of observations; similarly, the quantity on the right-hand side gives the frequency of occurrence of the pair $(n+k, n)$. It therefore follows that under equilibrium conditions, the probability, that in a given length of time we observe a transition from the state n to the state m is equal to the probability that (in an equal length of time) we observe a transition from the state m to the state n. It is precisely the symmetry between the past and the future which guarantees this equality between $H(n; m)$ and $H(m; n)$. A glance at Table VI shows that this is amply confirmed by observations. [It may be further noted that, in accordance with Eq. (419) the numbers in italics on the opposite sides of the principal diagonal are equal.] All this, is, of course, in entire agreement with Loschmidt's requirements.

On the other hand, it is also evident from Table VI, that after a relatively large number like 5, 6, or 7 a number much smaller, generally follows; in other words, the probability that a number $n(\gg\nu)$ will further increase on the next observation is very small indeed. This circumstance illustrates how molecular concentrations differing appreciably from the average value will *almost* always tend to change in the direction indicated on the macroscopic notions concerning diffusion [cf. Eq. (356)]. This corresponds exactly to one of Boltzmann's statements that the negative entropy curve almost always decreases from any point. However this may be, in course of time, an abnormal initial state will again recur as a consequence of fluctuations, and we shall now see how in spite of this possibility for recurrence, the *apparently* irreversible nature of the phenomenon comes into being.

(*b*) *The resolution of Zermelo's paradox.*—Let us first consider the case of intermittent observations. As we have already remarked in §2, the number 17 never occurred in one of Svedberg's sequences for which ν had the value 1.55. But the average time of recurrence for this state [according to Eq. (376)] is $10^{13}\tau$; and since $\tau=1/39$ min., for the sequence considered, $\Theta\sim 500{,}000$ years. Hence, the diffusion from the state $n=17$ will have all the *appearances* of an irreversible process simply because the average time of recurrence is so very long compared to the times during which the system is under observation.

Turning next to the case of continuous observations, we shall return to the example considered in §3. As we have seen (cf. Table VIII) the average time of recurrence of a state in which the number of molecules of oxygen contained in a sphere of radius $a \geqslant 5\times 10^{-5}$ cm (and $T=300°$K and $\nu=3\times 10^{19}$ cm^{-3}) will differ from the average value by 1 percent is very long indeed ($\Theta > 10^{68}$ seconds). The factor which is principally responsible for these large values for Θ is the exponential factor in Eq. (418). Accordingly, we may say, very roughly, that *the second law of thermodynamics is valid only for those diffusion processes in which the equalization of molecular concentrations which take place are by amounts appreciably greater than the root mean square relative fluctuation* (namely, $[\langle|n-\nu|^2\rangle_{Av}/\nu^2]^{\frac{1}{2}} = \nu^{-\frac{1}{2}}$). We have thus completely reconciled (at any rate, for the processes under discussion) the notion of irreversibility which is at the base of thermodynamics and the essential reversibility of all molecular phenomena demanded by statistical mechanics. This reconciliation has become possible only because we have been able to specify the limits of validity of the second law.

Quite generally, we may conclude with Smoluchowski that *a process appears irreversible* (*or reversible*) *according as whether the initial state is characterized by a long* (*or short*) *average time of recurrence compared to the times during which the system is under observation.*

5. The Effect of Gravity on the Brownian Motion: The Phenomenon of Sedimentation

The study of the effect of gravity on the Brownian motion provides an interesting illustration of the use to which Smoluchowski's equation [Eq. (312)]

$$(\partial w/\partial t) = \mathrm{div}_r\,(q\beta^{-2}\,\mathrm{grad}_r\, w - \mathbf{K}\beta^{-1}w) \tag{420}$$

can be put. In Eq. (420) **K** represents the acceleration caused by the external field of force. If the external field is that due to gravity, we can write

$$K_x = 0;\quad K_y = 0;\quad K_z = -(1-(\rho_0/\rho))g, \tag{421}$$

provided the coordinate system has been so chosen that the z axis is in the vertical direction. In Eq. (421), g denotes the value of gravity, ρ the density of the Brownian particle and $\rho_0(\leqslant\rho)$ that of the surrounding fluid. Hence, for the case (421), Eq. (420) becomes

$$(\partial w/\partial t) = (q/\beta^2)\nabla^2 w + (1-(\rho_0/\rho))(g/\beta)(\partial w/\partial z). \tag{422}$$

It is seen that Eq. (422) is of the same general form as Eq. (126). Accordingly, we can interpret the phenomenon described by Eq. (422) as a process of diffusion in which the number of particles crossing elements of area normal to x, y, and z directions, per unit area and per unit time, are given, respectively, by [cf. Eq. (127)]

$$-D(\partial w/\partial x),\quad -D(\partial w/\partial y), \tag{423}$$

and

$$-D(\partial w/\partial z) - cw, \tag{424}$$

where

$$D = (q/\beta^2) = (kT/m\beta);\quad c = (1-(\rho_0/\rho))(g/\beta). \tag{425}$$

Thus, while the diffusion in the (x, y) plane takes exactly as in the field free case, the situation in the z direction is modified. If we, therefore, limit ourselves to considering only the distribution in the z direction, of particles uniformly distributed in the (x, y) plane, the appropriate differential equation is

$$\frac{\partial w}{\partial t} = D\frac{\partial^2 w}{\partial z^2} + c\frac{\partial w}{\partial z}. \tag{426}$$

Let us now suppose that the particle is initially at a height z_0 measured from the bottom of the vessel containing the solution. Then, the probability of occurrence of the various values of z at later times will be governed by the solution of Eq. (426) which satisfies the boundary conditions

$$\left.\begin{aligned} &w\to\delta(z-z_0)\quad \text{as}\quad t\to 0,\\ &D(\partial w/\partial z)+cw=0\quad \text{at}\quad z=0\quad \text{for all}\quad t>0. \end{aligned}\right\} \tag{427}$$

The second of two foregoing boundary conditions arises from the requirement that no particle shall cross the plane $z=0$ representing the bottom of the vessel [cf. Eq. (424)].

To obtain the solution of Eq. (426) satisfying the boundary conditions (427), we first introduce the following transformation of the variable [cf. Eq. (128)]

$$w = U(z, t)\exp\left[-\frac{c}{2D}(z-z_0) - \frac{c^2}{4D}t\right]. \tag{428}$$

Equation (426) reduces to the standard form

$$(\partial U/\partial t) = D(\partial^2 U/\partial z^2) \tag{429}$$

while the boundary conditions (427) become

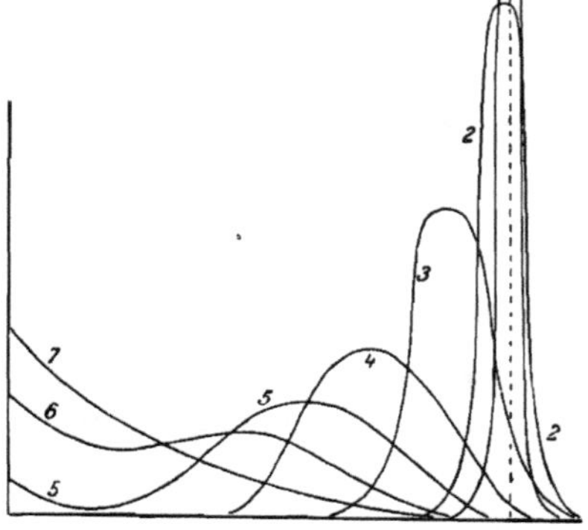

FIG. 7.

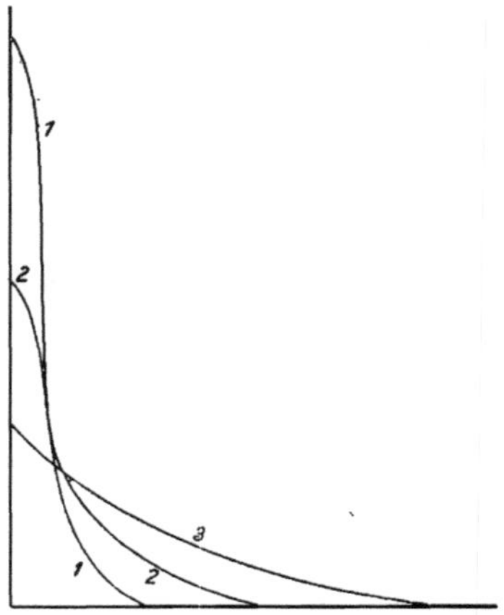

FIG. 8.

$$\left.\begin{array}{l} U\to\delta(z-z_0) \quad \text{as} \quad t\to 0, \\ D(\partial U/\partial z)+(1/2)cU=0 \quad \text{at} \quad z=0 \quad \text{for all} \quad t>0. \end{array}\right\} \tag{430}$$

Solving Eq. (429) with boundary conditions of the form (430) is a standard problem in the theory of heat conduction. We have

$$U=\frac{1}{2(\pi Dt)^{\frac{1}{2}}}\{\exp[-(z-z_0)^2/4Dt]+\exp[-(z+z_0)^2/4Dt]\} + \frac{c}{2D(\pi Dt)^{\frac{1}{2}}}\int_{z_0}^{\infty}\exp\left[-\frac{(\alpha+z)^2}{4Dt}+\frac{c(\alpha-z_0)}{2D}\right]d\alpha. \tag{431}$$

After some elementary transformations, Eq. (431) takes the form

$$U=\frac{1}{2(\pi Dt)^{\frac{1}{2}}}\{\exp[-(z-z_0)^2/4Dt]+\exp[-(z+z_0)^2/4Dt]\} + \frac{c}{D\sqrt{\pi}}\exp\left[\frac{c^2t}{4D}-\frac{c(z+z_0)}{2D}\right]\int_{\frac{z+z_0-ct}{2(Dt)^{\frac{1}{2}}}}^{\infty}\exp(-x^2)dx. \tag{432}$$

Returning to the variable w we have [cf. Eq. (428)]

$$w(t,z;z_0)=\frac{1}{2(\pi Dt)^{\frac{1}{2}}}\{\exp[-(z-z_0)^2/4Dt]+\exp[-(z+z_0)^2/4Dt]\} \times\exp\left[-\frac{c}{2D}(z-z_0)-\frac{c^2}{4D}t\right]+\frac{c}{D\sqrt{\pi}}e^{-cz/D}\int_{\frac{z+z_0-ct}{2(Dt)^{\frac{1}{2}}}}^{\infty}\exp(-x^2)dx \tag{433}$$

which is the required solution. In Fig. 7 we have illustrated according to Eq. (433) the distributions $w(z,t;z_0)$ for a given value of z_0 and various values of t.

If we suppose that at time $t=0$ we have a large number of particles distributed uniformly in the plane $z=z_0$ then in the first instance diffusion takes place as in the field free case (curves 1 and 2). However, gravity makes itself felt very soon (curves 3, 4, and 5) and the maximum begins to be displaced to lower values of z with the velocity c; at the same time, the maximum becomes flatter on account of the random motions experienced by the particles. Once the probability of finding particles near enough to the bottom of the vessel becomes appreciable, the curves again begin to rise upwards (curves 5 and 6) on account of the reflection which the particles suffer at $z=0$; and, finally as $t\rightarrow\infty$ we obtain the equilibrium distribution

$$w(z,\ \infty\ ;\ z_0)=(c/D)e^{-cz/D}. \tag{434}$$

Since [cf. Eq. (425)]

$$(c/D)=(1-(\rho_0/\rho))(mg/kT), \tag{435}$$

we see that the equilibrium distribution (434) represents simply the law of isothermal atmospheres in its standard form.

The example we have just considered provides a further illustration of a case to which the conventional notions concerning entropy and the second law of thermodynamics cannot be applied. For the state of maximum entropy for the system consisting of the Brownian particles and the surrounding fluid, is that in which all the particles are at $z=0$; and, on strict thermodynamical principles we should conclude that with the continued operation of dissipative forces like dynamical friction, the state of maximum entropy will be attained. But according to Eq. (434), as $t\rightarrow\infty$ though the state of maximum entropy $z=0$ has the maximum probability, it is *not* true that the average value of the height at which the particles will be found is also zero. Actually, for the equilibrium distribution (434), we have

$$\langle z\rangle_{\mathrm{Av}}=(D/c)=(kT/mg)[\rho/(\rho-\rho_0)], \tag{436}$$

which is the height of the equivalent homogeneous atmosphere. Moreover, even if the particles were initially at $z=0$, they will not continue to stay there. For, setting $z_0=0$ in Eq. (433) we find that

$$w(z,\ t;\ 0)=(1/(\pi Dt)^{\frac{1}{2}})\exp\left[-(z+ct)^2/4Dt\right]+(c/D\sqrt{\pi})e^{-cz/D}\int_{\frac{z-ct}{2(Dt)^{\frac{1}{2}}}}^{\infty}\exp(-x^2)dx. \tag{437}$$

Equation (437) shows that as $t\rightarrow\infty$ we are again led to the equilibrium distribution (434) (see Fig. 8). Hence, the particles do a certain amount of mechanical work *at the expense of the internal energy of the surrounding fluid;* this is of course contrary to the strict interpretation of the second law of thermodynamics. The average work done in this manner is given by [if we use Eq. (436)]

$$\langle A\rangle_{\mathrm{Av}}=m(1-(\rho_0/\rho))g\bar{z}=kT, \tag{438}$$

per particle. Hence, on the average there is a *decrease* in entropy of amount k per particle:

$$\langle S\rangle_{\mathrm{Av}}=S_{\max}-Nk, \tag{439}$$

where N denotes the number of Brownian particles. However, as Smoluchowski has pointed out, this work done at the expense of the internal energy of the surrounding fluid cannot be utilized to run a heat engine with an efficiency higher than that of the Carnot cycle.

We may further note that except for values of $z\lesssim D/c$, a particle has a greater probability to descend than it has to ascend. As $z\rightarrow 0$ the converse is true. We may therefore say that the tendency for the entropy to *increase* (almost always) for particles at $z\gg D/c$ is compensated by the tendency of the entropy to *decrease* for particles very near $z=0$; so that, on the average, a steady state is maintained. Of course, we have a finite probability for particles, occasionally to ascend to very great heights; but in accordance with the conclusions of §4 we should expect that the average time of recurrence for such abnormal states must be very long indeed.

6. The Theory of Coagulation in Colloids

Smoluchowski discovered a very interesting application of the theory of Brownian motion in the phenomenon of coagulation exhibited by colloidal particles when an electrolyte is added to the solution. Smoluchowski's theory of this phenomenon is based on a suggestion of Zsigmondy that coagulation results as a consequence of each colloidal particle being surrounded (on the addition of an electrolyte) by a *sphere of influence* of a certain radius R such that the Brownian motion of a particle proceeds unaffected only so long as no other particle comes within its sphere of influence and that when the particles do come within a distance R they stick to one another to form a single unit. We are not concerned here with the physico-chemical basis for Zsigmondy's suggestion except perhaps to remark that the spheres of influence are supposed to originate in the formation of electric double layers around each particle; we are here interested only in the application of the principles of Brownian motion which is possible on the acceptance of Zsigmondy's suggestion. However, we may formulate somewhat more explicitly the problem we wish to investigate:

We imagine that initially the colloidal solution contains only single particles all similar to one another and of the same spherical size. We now suppose that at time $t=0$ an (appropriate) electrolyte is added to the solution in such a way that the resulting electrolytic concentration is uniform throughout the solution. The particles are now supposed to be all instantaneously surrounded by spheres of influence of radius R. From this instant onwards, each particle will continue to describe the original Brownian motion only so long as no other particle comes within its sphere of influence. Once two particles do approach to within this distance R they will coalesce to form a "*double particle.*" This double particle will also describe Brownian motion but at a reduced rate consequent to its increased size. This double particle will, in turn, continue to remain as such only so long as it does not come within the appropriate spheres of influence of a single or another double particle: when this happens we shall have the formation of a triple or a quadruple particle; and, so on. The continuation of this process will eventually lead to the total coagulation of all the colloidal particles into one single mass.

The problem we wish to solve is the specification of the concentrations $\nu_1, \nu_2, \nu_3, \nu_4, \cdots$, of single, double, triple, quadruple, etc., particles at time t given that at time $t=0$ there were $\nu_0(=\nu_1[0])$ single particles.

As a preliminary to the discussion of the general problem formulated in the preceding paragraph we shall first consider the following more elementary situation:

A particle, assumed fixed in space, is in a medium of infinite extent in which a number of similar Brownian particles are distributed uniformly at time $t=0$. Further, if the stationary particle is assumed to be surrounded by a sphere of influence of radius R what is the rate at which particles arrive on the sphere of radius R surrounding the fixed particle?

We shall suppose that the stationary particle is at the origin of our system of coordinates. Then, in accordance with our definition of a sphere of influence, we can replace the surface $|\mathbf{r}|=R$ by a perfect absorber [cf. I, §5, see particularly Eq. (115)]. We have therefore to seek a solution of the diffusion equation [cf. Eqs. (173) and (306)]

$$(\partial w/\partial t)=D\nabla^2 w;\quad D=(q/\beta^2)=(kT/6\pi a\eta), \tag{440}$$

which satisfies the boundary conditions

$$\left.\begin{aligned} &w\equiv\nu=\text{constant, at}\quad t=0,\quad \text{for}\quad |\mathbf{r}|>R,\\ &w\equiv 0\quad \text{at}\quad |\mathbf{r}|=R\quad \text{for}\quad t>0.\end{aligned}\right\} \tag{441}$$

In the first of the two foregoing boundary conditions ν denotes the average concentration of the particles exterior to $|\mathbf{r}|=R$ at time $t=0$.

Since w can depend only on the distance r from the center, the form of the diffusion equation (440) appropriate to this case is

$$(\partial/\partial t)(rw)=D(\partial^2/\partial r^2)(rw). \tag{442}$$

The solution of this equation satisfying the boundary conditions (441) is

$$w=\nu\left[1-\frac{R}{r}+\frac{2R}{r\sqrt{\pi}}\int_0^{(r-R)/2(Dt)^{\frac{1}{2}}}\exp(-x^2)dx\right]. \tag{443}$$

From Eq. (443) it follows that the rate at which particles arrive at the surface $|\mathbf{r}|=R$ is given by [cf. Eq. (117)]

$$4\pi D\left(r^2\frac{\partial w}{\partial r}\right)_{r=R}=4\pi DR\nu\left(1+\frac{R}{(\pi Dt)^{\frac{1}{2}}}\right). \tag{444}$$

Equation (444) gives the rate at which particles describing Brownian motion will coalesce with a stationary particle surrounded by a sphere of influence of radius R. Suppose, now, that the particle we have assumed to be stationary is also describing Brownian motion. What is the corresponding generalization of (444)? In considering this generalization we shall not suppose that the diffusion coefficients characterizing the two particles which coalesce to form a multiple particle are necessarily the same. Under these circumstances we have clearly to deal with the *relative displacements* of the two particles; and it can be readily shown that the relative displacements between two particles describing Brownian motions independently of each other and with the diffusion coefficients D_1 and D_2 also follows the laws of Brownian motion with the diffusion coefficient $D_{12}=D_1+D_2$. For, the probability that the relative displacement of two particles, initially, together at $t=0$, lies between $\mathbf{r}$ and $\mathbf{r}+d\mathbf{r}$ is clearly

$$\left.\begin{aligned}W(\mathbf{r})d\mathbf{r}&=d\mathbf{r}\int_{-\infty}^{+\infty}W_1(\mathbf{r}_1)W_2(\mathbf{r}_1+\mathbf{r})d\mathbf{r}_1\\&=\frac{d\mathbf{r}}{(4\pi D_1t)^{\frac{3}{2}}(4\pi D_2t)^{\frac{3}{2}}}\int_{-\infty}^{+\infty}\exp(-|\mathbf{r}_1|^2/4D_1t)\exp(-|\mathbf{r}_1+\mathbf{r}|^2/4D_2t)d\mathbf{r}_1\end{aligned}\right\} \tag{445}$$

or, as may be readily verified [cf. the remarks following Eq. (62)]

$$W(\mathbf{r})=(1/[4\pi(D_1+D_2)t]^{\frac{3}{2}})\exp(-|\mathbf{r}|^2/4(D_1+D_2)t). \tag{446}$$

On comparing this distribution of the relative displacements with the corresponding result for the individual displacements [see for example Eq. (172)] we conclude that the relative displacements do follow the laws of Brownian motion with the diffusion coefficient (D_1+D_2).

Thus, the required generalization of Eq. (444) is

$$4\pi(D_1+D_2)R\nu\left(1+\frac{R}{[\pi(D_1+D_2)t]^{\frac{1}{2}}}\right). \tag{447}$$

More generally, let us consider two sorts of particles with concentrations ν_i and ν_k. Let the respective diffusion coefficients be D_i and D_k. Further, let R_{ik} denote the distance to which two particles (one of each sort) must approach in order that they may coalesce to form a multiple particle. Then, the rate of formation of the multiple particles by the coagulation of the particles of the kind considered is clearly given by

$$J_{i+k}dt=4\pi D_{ik}R_{ik}\nu_i\nu_k\left(1+\frac{R_{ik}}{(\pi D_{ik}t)^{\frac{1}{2}}}\right)dt \tag{448}$$

where we have written

$$D_{ik}=D_i+D_k. \tag{449}$$

In our further discussions, we shall ignore the second term in the parenthesis on the right-hand side of Eq. (447); this implies that we restrict ourselves to time intervals $\Delta t\gg R^2/D$. In most cases of practical interest, this is justifiable as $R^2/D \sim 10^{-3}-10^{-4}$ second. With this understanding we can write

$$J_{i+k}dt \simeq 4\pi D_{ik}R_{ik}\nu_i\nu_k dt. \tag{450}$$

Using Eq. (450) we can now write down the fundamental differential equations which govern the variations of $\nu_1, \nu_2, \cdots, \nu_k, \cdots$ (of single, double, $\cdots$, k-fold, $\cdots$,) particles with time:

Thus, considering the variation of the number of k-fold particles with time, we have in analogy with the equations of chemical kinetics

$$\frac{d\nu_k}{dt}=4\pi\left(\tfrac{1}{2}\sum_{i+j=k}\nu_i\nu_j D_{ij}R_{ij}-\nu_k\sum_{j=1}^{\infty}\nu_j D_{kj}R_{kj}\right) \quad (k=1, \cdots). \tag{451}$$

In this equation the first summation on the right-hand side represents the increase in ν_k due to the formation of k-fold particles by the coalescing of an i-fold and a j-fold particle (with $i+j=k$), while the second summation represents the decrease in ν_k due to the formation of $(k+j)$-fold particles in which one of the interacting particles is k-fold.

A general solution of the infinite system of Eq. (451) which will be valid under all circumstances does not seem feasible. But a special case considered by Smoluchowski appears sufficiently illustrative of the general solution.

First, concerning R_{ik}, the assumption is made that

$$R_{ik}=\tfrac{1}{2}(R_i+R_k), \tag{452}$$

where R_i and R_k are the radii of the spheres of influence of the i-fold and the k-fold particles. We can, if we choose, regard the assumption (452) as equivalent to Zsigmondy's suggestion concerning the basic cause of coagulation.

Again, according to Eq. (440), the diffusion coefficient is inversely proportional to the radius of the particle; and on the basis of experimental evidence it appears that the radii of the spheres of influence of various multiple particles are proportional to the radii of the respective particles. We therefore make the additional assumption that

$$D_iR_i=DR \quad (i=1, \cdots), \tag{453}$$

where D and R denote, respectively, the diffusion coefficient and the radius of the sphere of influence of the single particles.

Combining Eqs. (449), (452), and (453) we have

$$D_{ik}R_{ik}=\tfrac{1}{4}(D_i+D_k)(R_i+R_k)=\tfrac{1}{4}DR(R_i^{-1}+R_k^{-1})(R_i+R_k)=\tfrac{1}{4}DR(R_i+R_k)^2R_i^{-1}R_k^{-1}. \tag{454}$$

Finally, for the sake of mathematical simplicity we make the (not very plausible) assumption that

$$R_i=R_k. \tag{455}$$

Thus, with all these assumptions

$$D_{ik}R_{ik}=2DR. \tag{456}$$

In view of (456), Eq. (451) becomes

$$\frac{d\nu_k}{dt}=8\pi DR\left(\tfrac{1}{2}\sum_{i+j=k}\nu_i\nu_j-\nu_k\sum_{j=1}^{\infty}\nu_j\right) \quad (k=1, \cdots). \tag{457}$$

If we now let

$$\tau=4\pi DRt, \tag{458}$$

Eq. (457) takes the more convenient form

$$\frac{d\nu_k}{d\tau}=\sum_{i+j=k}\nu_i\nu_j-2\nu_k\sum_{j=1}^{\infty}\nu_j \quad (k=1, \cdots). \tag{459}$$

From Eq. (459) we readily find that

$$\left.\begin{aligned}\frac{d}{dt}\Big(\sum_{k=1}^{\infty}\nu_k\Big)&=\sum_{i=1}^{\infty}\sum_{j=1}^{\infty}\nu_i\nu_j-2\sum_{k=1}^{\infty}\sum_{j=1}^{\infty}\nu_k\nu_j,\\ &=-\Big(\sum_{k=1}^{\infty}\nu_k\Big)^2,\end{aligned}\right\}\tag{460}$$

or,

$$\sum_{k=1}^{\infty}\nu_k=\frac{\nu_0}{1+\nu_0\tau},\tag{461}$$

remembering that at $t=0$, $\sum\nu_k=\nu_0$.

Using the integral (461) we can successively obtain the solutions for ν_1, ν_2, etc. Thus, considering the equation for ν_1 we have [cf. Eq. (459)]

$$d\nu_1/dt=-2\nu_1\sum_{k=1}^{\infty}\nu_k=-2\nu_1\nu_0/(1+\nu_0\tau);\tag{462}$$

in other words,

$$\nu_1=\frac{\nu_0}{(1+\nu_0\tau)^2},\tag{463}$$

again using the boundary condition that $\nu_1=\nu_0$ at $t=0$. Proceeding in this manner we can prove (by induction) that

$$\nu_k=\nu_0[(\nu_0\tau)^{k-1}/(1+\nu_0\tau)^{k+1}]\quad(k=1,\,2,\,\cdots).\tag{464}$$

In Fig. 9 we have illustrated the variations of $\sum\nu_k$, ν_1, ν_2, $\cdots$ with time. We shall not go into the details of the comparison of the predictions of this theory with the data of observations. Such comparisons have been made by Zsigmondy and others and the general conclusion is that Smoluchowski's theory gives a fairly satisfactory account of the broad features of the coagulation phenomenon.

7. The Escape of Particles over Potential Barriers

As a final illustration of the application of the principles of Brownian motion we shall consider, following Kramers, the problem of the escape of particles over potential barriers. The solution to this problem has important bearings on a variety of physical, chemical, and astronomical problems.

The situation we have in view is the following:

Limiting ourselves for the sake of simplicity to a one-dimensional problem, we consider a particle moving in a potential field $\mathfrak{V}(x)$ of the type shown in Fig. 10; more generally, we may consider an

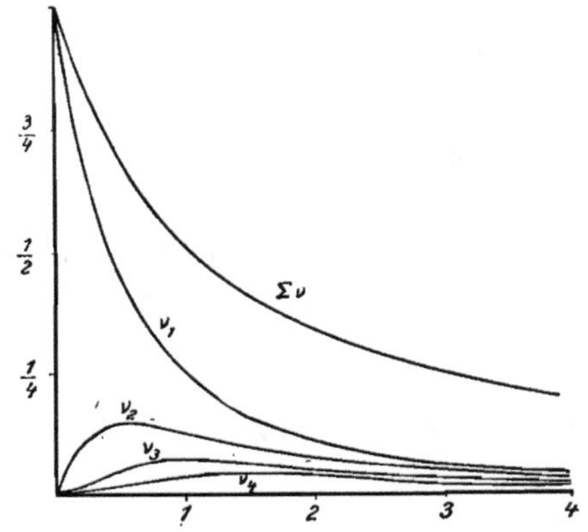

FIG. 9.

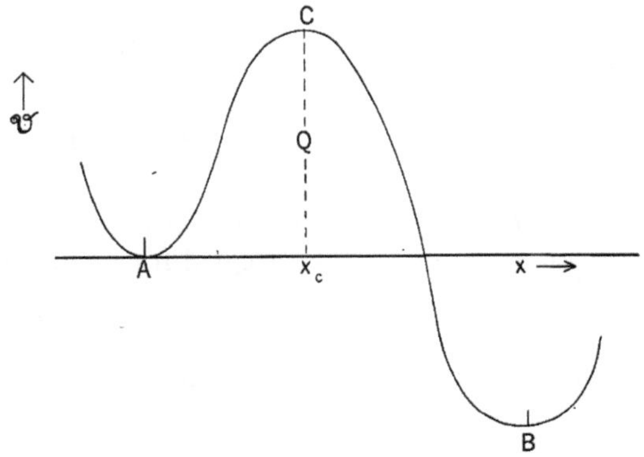

FIG. 10.

ensemble of particles moving in the potential field $\mathfrak{V}(x)$ without any mutual interference. We suppose that the particles are initially caught in the potential hole at A. The general problem we wish to solve concerns the rate at which particles will escape over the potential barrier in consequence of Brownian motion.

In the most general form, the solution to the problem formulated in the foregoing paragraph is likely to be beset with considerable difficulties. But a special case of interest arises when the height of the potential barrier is large compared to the energy of the thermal motions:

$$mQ \gg kT. \tag{465}$$

Under these circumstances, the problem can be treated as one in which the conditions are *quasi-stationary*. More specifically, we may suppose that to a high degree of accuracy a Maxwell-Boltzmann distribution obtains in the neighborhood of A. But the equilibrium distribution will not obtain for all values of x. For, by assumption, the density of particles beyond C is very small compared to the equilibrium values; and in consequence of this there will be a slow diffusion of particles (across C) tending to restore equilibrium conditions throughout. If the barrier were sufficiently high, this diffusion will take place as though stationary conditions prevailed.

Assuming first that we are interested only in time intervals that are long compared to the time of relaxation β^{-1} we can use Smoluchowski's Eq. (312). Under stationary conditions, Smoluchowski's equation predicts a current density j given by [cf. Eq. (316)]

$$\mathbf{j}\cdot\int_A^B \beta e^{m\mathfrak{V}/kT}d\mathbf{s} = (kT/m)we^{m\mathfrak{V}/kT}\Big|_B^A, \tag{466}$$

where, in the integral on the right-hand side, the path of integration of $\mathbf{s}$ from A to B is arbitrary. In our present case β is a constant and, since further we are dealing with a one-dimensional problem, we can express Eq. (466) in the form

$$j = \frac{kT}{m\beta}\frac{we^{m\mathfrak{V}(x)/kT}\Big|_B^A}{\int_A^B e^{m\mathfrak{V}(x)/kT}dx}. \tag{467}$$

Now, the number of particles ν_A in the vicinity of A can be calculated; for, in accordance with our earlier remarks we shall be justified in assuming that the Maxwell-Boltzmann distribution

$$d\nu_A = w_A e^{-m\mathfrak{V}/kT}dx \tag{468}$$

is valid in the neighborhood of A. If we now further suppose that

$$\mathfrak{V} \simeq \tfrac{1}{2}\omega_A^2 x^2 \quad (\omega_A = \text{constant};\ x \sim 0), \tag{469}$$

we obtain from Eq. (468)

$$\nu_A = w_A \int_{-\infty}^{+\infty} \exp\left(-m\omega_A^2 x^2/2kT\right)dx, \tag{470}$$

where the range of integration over x has been extended from $-\infty$ to $+\infty$ in view of the fact that the main contribution to the integral for ν_A must arise only from a small region near $x=0$. Hence,

$$\nu_A = (w_A/\omega_A)(2\pi kT/m)^{\frac{1}{2}}. \tag{471}$$

Returning to Eq. (467), we can write with sufficient accuracy [cf. Eq. (469)]:

$$j \simeq \frac{kT}{m\beta} w_A \left\{\int_A^B e^{m\mathfrak{V}/kT}dx\right\}^{-1}. \tag{472}$$

In writing Eq. (472) we have assumed that the density of particles near B is very small: this is true to begin with anyway.

From Eqs. (471) and (472) we directly obtain for the rate at which a particle, initially caught in the potential hole at A, will escape over the barrier at C, the expression

$$P=\frac{j}{\nu_A}=\frac{\omega_A}{\beta}\left(\frac{kT}{2\pi m}\right)^{\frac{1}{2}}\left\{\int_A^B e^{m\mathfrak{B}/kT}dx\right\}^{-1}. \tag{473}$$

The principal contribution to the integral in the curly brackets in the foregoing equation arises from only a very small region near C [on account of the strong inequality (465)]. The value of the integral will therefore depend, very largely, only on the shape of the potential curve in the immediate neighborhood of C. If we now suppose that near $x=x_C$, $\mathfrak{B}(x)$ has a continuous curvature, we may write

$$\mathfrak{B}\simeq Q-\tfrac{1}{2}\omega_C^2(x-x_C)^2 \quad (\omega_C=\text{constant};\ x\sim x_C). \tag{474}$$

On this assumption, to a sufficient degree of accuracy we have

$$\left.\begin{aligned}\int_A^B e^{m\mathfrak{B}/kT}dx &\simeq e^{mQ/kT}\int_{-\infty}^{+\infty}\exp\left[-m\omega_C^2(x-x_C)^2/2kT\right]dx,\\ &= e^{mQ/kT}(2\pi kT/m\omega_C^2)^{\frac{1}{2}}.\end{aligned}\right\} \tag{475}$$

Combining Eqs. (473) and (475) we obtain

$$P=(\omega_A\omega_C/2\pi\beta)e^{-mQ/kT}, \tag{476}$$

which gives the probability, per unit time, that a particle originally in the potential hole at A, will escape to B crossing the barrier at C.

The formula (476) has been derived on the basis of Eq. (467) and this implies, as we have already remarked, that we are ignoring effects which take place in intervals of the order β^{-1}. Alternatively, we may say that the validity of Eq. (476) depends on how large the coefficient of dynamical friction β is: if β were sufficiently large, the formula (476) for P may be expected to provide an adequate approximation [see Eqs. (507) and (508) below]. On the other hand, if this should not be the case, we must, in accordance with our remarks in Chapter II, §4, subsection (*vi*), base our discussion of the generalized Liouville Eq. (249) in phase space; and in one dimension this equation has the form

$$\frac{\partial W}{\partial t}+u\frac{\partial W}{\partial x}+K\frac{\partial W}{\partial u}=\beta u\frac{\partial W}{\partial u}+\beta W+q\frac{\partial^2 W}{\partial u^2}, \tag{477}$$

where it may be recalled that

$$q=\beta(kT/m);\quad K=-(\partial\mathfrak{B}/\partial x). \tag{478}$$

In II, §5 we have shown that the Maxwell-Boltzmann distribution identically satisfies Eq. (249). Accordingly,

$$W=C\exp\left[-(mu^2+2m\mathfrak{B})/2kT\right], \tag{479}$$

where C is a constant, satisfies Eq. (477). However, under the conditions of our problem the equilibrium distribution (479) cannot be valid for *all* values of x; for, if it were, there would be no diffusion across the barrier at C and the conditions of the problem would not be met. On the other hand, we do expect the distribution (479) to be realized to a high degree of accuracy in the neighborhood of A. We, therefore, look for a stationary solution of Eq. (477) of the form

$$W=CF(x,u)\exp\left[-m(u^2+2\mathfrak{B})/2kT\right], \tag{480}$$

where $F(x, u)$ is very nearly unity in the neighborhood of $x=0$. Since we have further supposed that the density of particles in the region B is quite negligible, we should also require that $F(x, u)\to 0$ for values of x appreciably greater than $x=x_C$. We may express these conditions formally in the form

$$\begin{aligned} F(x, u)&\simeq 1 \quad \text{at} \quad x\sim 0, \\ F(x, u)&\simeq 0 \quad \text{for} \quad x\gg x_C. \end{aligned} \tag{481}$$

We shall now show how such a function $F(x, u)$ can be determined.

First of all it is evident that for the purposes of determining the rate of escape of particles across the barrier at C it is particularly important to determine F accurately in this region. Assuming that in the vicinity of C, $\mathfrak{B}$ has the form (474) and that stationary conditions prevail throughout, the equation for W in the neighborhood of $x=x_C$ becomes [cf. Eq. (477)]:

$$u\frac{\partial W}{\partial X}+\omega_C^2X\frac{\partial W}{\partial u}=\beta u\frac{\partial W}{\partial u}+\beta W+q\frac{\partial^2 W}{\partial u^2}, \tag{482}$$

where for the sake of brevity, we have used

$$X=x-x_C. \tag{483}$$

According to Eqs. (474), (480), and (483) the appropriate form for W valid in the region C, is

$$W=Ce^{-mQ/kT}F(X, u)\exp\left[-m(u^2-\omega_C^2X^2)/2kT\right]. \tag{484}$$

Substituting for W according to this equation in Eq. (482), we obtain

$$u\frac{\partial F}{\partial X}+\omega_C^2X\frac{\partial F}{\partial u}=q\frac{\partial^2 F}{\partial u^2}-\beta u\frac{\partial F}{\partial u}. \tag{485}$$

It is seen that $F=$constant satisfies this equation identically: this solution corresponds of course to the equilibrium distribution. However, the solution of Eq. (485) which we are seeking must satisfy the boundary conditions [cf. Eq. (481)]

$$\begin{aligned} F(X, u)&\to 1 \quad \text{as} \quad X\to -\infty, \\ F(X, u)&\to 0 \quad \text{as} \quad X\to +\infty. \end{aligned} \tag{486}$$

Assume for F the form

$$F\equiv F(u-aX)=F(\xi) \quad \text{(say)}, \tag{487}$$

where a is, for the present, an unspecified constant. Substituting this form of F in Eq. (485) we obtain

$$-\left[(a-\beta)u-\omega_C^2X\right]\frac{dF}{d\xi}=q\frac{d^2F}{d\xi^2}. \tag{488}$$

In order that Eq. (488) be consistent it is clearly necessary that [cf. Eq. (487)]

$$\left[\omega_C^2/(a-\beta)\right]=a; \tag{489}$$

and in this case Eq. (488) becomes

$$-(a-\beta)\xi\frac{dF}{d\xi}=q\frac{d^2F}{d\xi^2}. \tag{490}$$

Equation (490) is readily integrated to give

$$F=F_0\int^{\xi}\exp\left[-(a-\beta)\xi^2/2q\right]d\xi, \tag{491}$$

where F_0 is a constant. On the other hand, according to Eq. (489) a is the root of the equation

$$a^2 - a\beta - \omega_C^2 = 0; \tag{492}$$

i.e.,

$$a = (\beta/2) \pm ([\beta^2/4] + \omega_C^2)^{\frac{1}{2}}. \tag{493}$$

If we choose for a the *positive root*, then

$$a - \beta = ([\beta^2/4] + \omega_C^2)^{\frac{1}{2}} - (\beta/2) \tag{494}$$

is also positive, and as we shall show presently, the solution (491) leads to an F which satisfies the required boundary conditions (486). For, by choosing

$$F_0 = [(a-\beta)/2\pi q]^{\frac{1}{2}}, \tag{495}$$

and setting the lower limit of integration in Eq. (491) as $-\infty$ we obtain the solution

$$F = \left(\frac{a-\beta}{2\pi q}\right)^{\frac{1}{2}} \int_{-\infty}^{\xi} \exp\,[-(a-\beta)\xi^2/2q]d\xi, \tag{496}$$

which satisfies the conditions

$$F \to 1 \quad \text{as} \quad \xi \to +\infty; \quad F \to 0 \quad \text{as} \quad \xi \to -\infty. \tag{497}$$

On the other hand, since $\xi = u - aX$ and $a(=[(\beta/2)^2 + \omega_C^2]^{\frac{1}{2}} + [\beta/2])$ is positive, $\xi \to +\infty$ or $-\infty$ is the same as $X \to -\infty$ or $+\infty$; in other words, the solution (496) for F satisfies the necessary boundary conditions (486).

Combining Eqs. (484) and (496) we have, therefore, the solution

$$W = C[(a-\beta)/2\pi q]^{\frac{1}{2}} e^{-mQ/kT} \exp\,[-m(u^2 - \omega_C^2 X^2)/2kT] \int_{-\infty}^{\xi} \exp\,[-(a-\beta)\xi^2/2q]d\xi. \tag{498}$$

Equation (498) is, of course, valid only in the neighborhood of C.

In the vicinity of A we have the solution [cf. Eqs. (469) and (479)]

$$W = C\exp\,[-m(u^2 + \omega_A^2 x^2)/2kT]. \tag{499}$$

Accordingly, the number of particles, ν_A, in the potential hole at A is given by

$$\left.\begin{aligned} \nu_A &\simeq C\int_{-\infty}^{+\infty}\int_{-\infty}^{+\infty} \exp\,[-m(u^2 + \omega_A^2 x^2)/2kT]dxdu, \\ &= C(2\pi kT/m\omega_A). \end{aligned}\right\} \tag{500}$$

(This equation will enable us to normalize the distribution in such a way so as to correspond to one particle in the potential hole: for this purpose we need only choose $C = m\omega_A/2\pi kT$.)

Now, the diffusion current across C is given by

$$j = \int_{-\infty}^{+\infty} W(X=0;\, u)udu, \tag{501}$$

or, using the solution (498), we have

$$j = C[(a-\beta)/2\pi q]^{\frac{1}{2}} e^{-mQ/kT} \int_{-\infty}^{+\infty} duu \exp\,(-mu^2/2kT) \int_{-\infty}^{u} d\xi \exp\,[-(a-\beta)\xi^2/2q]. \tag{502}$$

After an integration by parts, we find

$$j = C[(a-\beta)/2\pi q]^{\frac{1}{2}} (kT/m) e^{-mQ/kT} \int_{-\infty}^{+\infty} \exp\,\{-u^2[m/2kT + (a-\beta)/2q]\}du. \tag{503}$$

But [cf. Eq. (478)]

$$(m/2kT)+[(a-\beta)/2q]=(a/2q). \tag{504}$$

From Eqs. (503) and (504) we now obtain

$$j=C(kT/m)[(a-\beta)/a]^{\frac{1}{2}}e^{-mQ/kT}. \tag{505}$$

Hence, the rate of escape of particles across C is given by

$$P=(j/\nu_A)=(\omega_A/2\pi)[(a-\beta)/a]^{\frac{1}{2}}e^{-mQ/kT}, \tag{506}$$

or, substituting for a and $a-\beta$ according to Eqs. (493) and (494), we find after some elementary reductions, that

$$P=(\omega_A/2\pi\omega_C)([\beta^2/4+\omega_C^2]^{\frac{1}{2}}-[\beta/2])e^{-mQ/kT}. \tag{507}$$

If

$$\beta\gg 2\omega_C \tag{508}$$

we readily verify that our present "exact" formula for P reduces to our earlier result (476) derived on the basis of the Smoluchowski equation. But (507) now provides in addition the precise condition for the approximate validity of (476). On the other hand, for $\beta\rightarrow 0$ we have

$$P=(\omega_A/2\pi)e^{-mQ/kT}\quad(\beta\rightarrow 0). \tag{509}$$

This last formula for P valid in the limit of vanishing dynamical friction, corresponds to what is sometimes called the approximation of the *transition-state method.*

CHAPTER IV

PROBABILITY METHODS IN STELLAR DYNAMICS: THE STATISTICS OF THE GRAVITATIONAL FIELD ARISING FROM A RANDOM DISTRIBUTION OF STARS

1. Fluctuations in the Force Acting on a Star; The Outline of the Statistical Method

One of the principal problems of stellar dynamics is concerned with the analysis of the nature of the force acting on a star which is a member of a stellar system. In a general way, it appears that we may broadly distinguish between the influence of the system as a whole and the influence of the immediate local neighborhood; the former will be a smoothly varying function of position and time while the latter will be subject to relatively rapid fluctuations (see below).

Considering first the influence of the system as a whole, it appears that we can express it in terms of the gravitational potential $\mathfrak{B}(\boldsymbol{r};t)$ derived from the density function $n(\boldsymbol{r},M;t)$ which governs the average spatial distribution of the stars of different masses at time t. Thus,

$$\mathfrak{B}(\boldsymbol{r};t)=-G\int_{-\infty}^{+\infty}\int_{0}^{\infty}\frac{Mn(\boldsymbol{r}_1,M;t)}{|\boldsymbol{r}_1-\boldsymbol{r}|}dMd\boldsymbol{r}_1, \tag{510}$$

where G denotes the constant of gravitation. The potential $\mathfrak{B}(\boldsymbol{r};t)$ derived in this manner may be said to represent the "smoothed out" distribution of matter in the stellar system. The force per unit mass acting on a star due to the "system as a whole" is therefore given by

$$\boldsymbol{K}=-\text{grad}\,\mathfrak{B}(\boldsymbol{r};t). \tag{511}$$

However, the fluctuations in the *complexion* of the local stellar distribution will make the instantaneous force acting on a star deviate from the value given by Eq. (511). To elucidate the nature and origin of these fluctuations, we surround the star under consideration by an element of volume σ, which we may suppose is small enough to contain, on the average, only a relatively few stars. The actual number of stars, which will be found in σ at any given instant, will not in general be the average number that will be expected to be in it, namely σn; it will be subject to fluctuations. These fluctuations will naturally be governed by a Poisson distribution with the variance σn [see Eq. (333)]. It is in direct consequence of this changing complexion of the local stellar distribution that the influence of the near neighbors on a star is variable. The average period of such a fluctuation is readily estimated: for the order of magnitude of the time involved is evidently that required for two stars to separate by a distance equal to the average distance D between the stars (see Appendix VII). We may, therefore, expect that the influence of the immediate neighborhood will fluctuate with an average period of the order of

$$T \simeq (D/(\langle |V|^2\rangle_{Av})^{\frac{1}{2}}); \qquad (512)$$

where $\langle |V|^2\rangle_{Av}^{\frac{1}{2}}$ denotes the root mean square relative velocity between two stars.

In the neighborhood of the sun, $D \sim 3$ parsecs, $\langle |V|^2\rangle_{Av}^{\frac{1}{2}} \sim 50$ km/sec. Hence

$$T \text{ (near the sun)} \simeq 6\times 10^4 \text{ years.} \qquad (513)$$

When we compare this time with the period of galactic rotation (which is about 2×10^8 years) we observe that in conformity with our earlier remarks, the fluctuations in the force acting on a star due to the changing local stellar distribution do occur with extreme rapidity compared to the rate at which any of the other physical parameters change. Accordingly we may write for the force per unit mass acting on a star, the expression

$$\mathfrak{F} = \boldsymbol{K}(\boldsymbol{r};t) + \boldsymbol{F}(t), \qquad (514)$$

where $\boldsymbol{K}$ is derived from the smoothed out distribution [as in Eqs. (510) and (511)] and $\boldsymbol{F}$ denotes the fluctuating force due to the near neighbors. Moreover, if Δt denotes an interval of time long compared to (512), we may write

$$\mathfrak{F}\Delta t = \boldsymbol{K}\Delta t + \boldsymbol{\delta}(t+\Delta t;t), \qquad (515)$$

where

$$\boldsymbol{\delta}(t+\Delta t;t) = \int_t^{t+\Delta t} \boldsymbol{F}(\xi)d\xi \quad (\Delta t \gg T). \qquad (516)$$

Under the circumstances stated (namely, $\Delta t \gg T$) the accelerations $\boldsymbol{\delta}(t+\Delta t;t)$ and $\boldsymbol{\delta}(t+2\Delta t;t+\Delta t)$ suffered during two successive intervals $(t+\Delta t, t)$ and $(t+2\Delta t, t+\Delta t)$ will not be expected to show any correlation. We may, therefore, anticipate the existence of a definite law of distribution which will govern the probability of occurrence of the different values of $\boldsymbol{\delta}(t+\Delta t;t)$. We thus see that the acceleration which a star suffers during an interval $\Delta t \gg T$ can be formally expressed as the sum of two terms: a *systematic* term $\boldsymbol{K}\Delta t$ due to the action of the gravitational field of the smoothed out distribution, and a *stochastic* term $\boldsymbol{\delta}(t+\Delta t;t)$ representing the influence of the near neighbors. Stated in this fashion, we recognize the similarity[8] between our present problems in stellar dynamics and those in the theory of Brownian motion considered in Chapters II and III. One important difference should however be noted: Under our present circumstances it is possible, as we shall presently see, to undertake an analysis of the statistical properties of $\boldsymbol{F}(t)$ and $\boldsymbol{\delta}(t+\Delta t;t)$ based on first principles and without appealing to any "intuitive" or *a priori* considerations as in the discussions of Brownian motion [see the remarks at the end of II, §1 and also those following Eq. (318)].

We shall now outline a general method which appears suitable for analyzing the statistical properties of $\boldsymbol{F}$.

The force $\boldsymbol{F}$ acting on a star, per unit mass, is given by

$$\boldsymbol{F} = G\sum_i \frac{M_i}{|\boldsymbol{r}_i|^3}\boldsymbol{r}_i, \qquad (517)$$

where M_i denotes the mass of a typical "field" star and $\boldsymbol{r}_i$ its position vector relative to the star under consideration; further, in Eq. (517) the summation is to be extended over all the neighboring stars. The actual value of $\boldsymbol{F}$ given by Eq. (517) at any particular instant of time will depend on the instantaneous complexion of the local stellar distribution; it is in consequence subject to fluctuations. We can therefore ask only for the probability of occurrence,

$$W(\boldsymbol{F})dF_xdF_ydF_z = W(\boldsymbol{F})d\boldsymbol{F}, \qquad (518)$$

of $\boldsymbol{F}$ in the range $\boldsymbol{F}$ and $\boldsymbol{F}+d\boldsymbol{F}$. In evaluating this probability distribution, we shall (consistent with the physical situations we have in view) suppose that fluctuations subject only to the restriction of a constant average density occur.

The probability distribution $W(\boldsymbol{F})$ of $\boldsymbol{F}$ can be obtained by a direct application of Markoff's method outlined in Chapter I, §3. We shall obtain the explicit form of this distribution (sometimes called the Holtsmark distribution) in §2 below, but we should draw attention, already at this stage, to the fact that the specification of $W(\boldsymbol{F})$ does *not* provide us with all the

[8] Cf. particularly Eq. (317) and Eq. (515) above.

necessary information concerning the fluctuating force $\boldsymbol{F}$ for an equally important aspect of $\boldsymbol{F}$ concerns the *speed of fluctuations.*

According to Eq. (517) the rate of change of $\boldsymbol{F}$ with time is given by

$$f=\frac{d\boldsymbol{F}}{dt}=G\sum_i M_i\left\{\frac{\boldsymbol{V}_i}{|\boldsymbol{r}_i|^3}-3\boldsymbol{r}_i\frac{(\boldsymbol{r}_i\cdot\boldsymbol{V}_i)}{|\boldsymbol{r}_i|^5}\right\}, \quad (519)$$

where $\boldsymbol{V}_i$ denotes the velocity of a typical field star *relative* to the star under consideration. It is now clear that the speed of fluctuations in $\boldsymbol{F}$ can be specified in terms of the bivariate distribution

$$W(\boldsymbol{F},\boldsymbol{f}) \quad (520)$$

which governs the probability of the simultaneous occurrence of prescribed values for both $\boldsymbol{F}$ and $\boldsymbol{f}$. It is seen that this distribution function $W(\boldsymbol{F},\boldsymbol{f})$ will depend on the assignment of *a priori* probability in the *phase space* in contrast to the distribution $W(\boldsymbol{F})$ of $\boldsymbol{F}$ which depends only on a similar assignment in the *configuration space.* Again, it is possible by an application of Markoff's method *formally* to write down a general expression for $W(\boldsymbol{F},\boldsymbol{f})$; but it does not appear feasible to obtain the required distribution function in an explicit form. However, as Chandrasekhar and von Neumann have shown, explicit formulae for *all* the first and the second moments of $\boldsymbol{f}$ for a given $\boldsymbol{F}$ can be obtained; and it appears possible to make some progress in the specification of the statistical properties of $\boldsymbol{F}$ in terms of these moments.

2. The Holtsmark distribution $W(F)$

We shall now obtain the stationary distribution $W(\boldsymbol{F})$ of the force $\boldsymbol{F}$ acting on a star, per unit mass, due to the gravitational attraction of the neighboring stars.

Without loss of generality we can suppose that the star under consideration is at the origin O of our system of coordinates. About O describe a sphere of radius R and containing N stars. In the first instance we shall suppose that

$$\boldsymbol{F}=G\sum_{i=1}^{N}\frac{M_i}{|\boldsymbol{r}_i|^3}\boldsymbol{r}_i=\sum_{i=1}^{N}\boldsymbol{F}_i. \quad (521)$$

But we shall subsequently let R and N tend to infinity simultaneously in such a way that

$$(4/3)\pi R^3 n=N$$
$$(R\to\infty\,;\ N\to\infty\,;\ n=\text{constant}). \quad (522)$$

This limiting process is permissible, in view of what we shall later show to be the case, namely, that the dominant contribution to $\boldsymbol{F}$ is made by the nearest neighbor [cf. Eqs. (560) and (564) below]; consequently, the formal extrapolation to infinity of the density of stars obtaining only in a given region of a stellar system can hardly affect the results to any appreciable extent.

Considering first the distribution $W_N(\boldsymbol{F})$ at the center of a finite sphere of radius R and containing N stars, we seek the probability that

$$\boldsymbol{F}_0 \leqslant \boldsymbol{F} \leqslant \boldsymbol{F}_0+d\boldsymbol{F}_0. \quad (523)$$

Applying Markoff's method to this problem we have [cf. Eqs. (51) and (52)]

$$W_N(\boldsymbol{F}_0)=\frac{1}{8\pi^3}\int_{-\infty}^{+\infty}\exp\left(-i\boldsymbol{\varrho}\cdot\boldsymbol{F}_0\right)A_N(\boldsymbol{\varrho})d\boldsymbol{\varrho}, \quad (524)$$

where

$$A_N(\boldsymbol{\varrho})=\prod_{i=1}^{N}\int_{M_i=0}^{\infty}\int_{|\boldsymbol{r}_i|=0}^{R}\exp\left(i\boldsymbol{\varrho}\cdot\boldsymbol{F}_i\right)\times\tau_i(\boldsymbol{r}_i,M_i)d\boldsymbol{r}_i dM_i. \quad (525)$$

In Eq. (525) $\tau_i(\boldsymbol{r}_i, M_i)$ governs the probability of occurrence of the ith star at the position $\boldsymbol{r}_i$ with a mass M_i. If we now suppose that only fluctuations which are compatible with a constant average density occur, then

$$\tau_i(\boldsymbol{r}_i,M_i)=(3/4\pi R^3)\tau(M), \quad (526)$$

where $\tau(M)$ now governs the frequency of occurrence of the different masses among the stars. With the assumption (526) concerning the τ_i's Eq. (525) reduces to

$$A_N(\boldsymbol{\varrho})=\left[\frac{3}{4\pi R^3}\int_{M=0}^{\infty}\int_{|\boldsymbol{r}|=0}^{R}\exp\left(i\boldsymbol{\varrho}\cdot\boldsymbol{\phi}\right)\times\tau(M)d\boldsymbol{r}dM\right]^N, \quad (527)$$

where we have written

$$\boldsymbol{\phi}=GM\boldsymbol{r}/|\boldsymbol{r}|^3. \quad (528)$$

We now let R and N tend to infinity according to Eq. (522). We thus obtain

$$W(F)=\frac{1}{8\pi^3}\int_{-\infty}^{+\infty}\exp(-i\varrho\cdot F)A(\varrho)d\varrho, \quad (529)$$

where

$$A(\varrho)=\lim_{R\to\infty}\left[\frac{3}{4\pi R^3}\int_{M=0}^{\infty}\int_{|r|=0}^{R}\exp(i\varrho\cdot\phi)\times\tau(M)drdM\right]^{4\pi R^3 n/3}. \quad (530)$$

Since,

$$\frac{3}{4\pi R^3}\int_{M=0}^{\infty}\int_{|r|=0}^{R}\tau(M)dMdr=1, \quad (531)$$

we can rewrite our expression for $A(\rho)$ in the form

$$A(\varrho)=\lim_{R\to\infty}\left[1-\frac{3}{4\pi R^3}\int_{=0}^{\infty}\int_{|r|=0}^{R}\tau(M)\times[1-\exp(i\varrho\cdot\phi)]drdM\right]^{4\pi R^3 n/3}. \quad (532)$$

The integral over r which occurs in Eq. (532) is seen to be absolutely convergent when extended over *all* $|r|$, i.e., also for $|r|\to\infty$. We can accordingly write

$$A(\varrho)=\lim_{R\to\infty}\left[1-\frac{3}{4\pi R^3}\int_{M=0}^{\infty}\int_{|r|=0}^{\infty}\tau(M)\times[1-\exp(i\varrho\cdot\phi)]drdM\right]^{4\pi R^3 n/3}, \quad (533)$$

or

$$A(\varrho)=\exp[-nC(\varrho)], \quad (534)$$

where

$$C(\varrho)=\int_{M=0}^{\infty}\int_{|r|=0}^{\infty}\tau(M)[1-\exp(i\varrho\cdot\phi)]drdM. \quad (535)$$

In the integral defining $C(\varrho)$ we shall introduce ϕ as the variable of integration instead of r. We readily verify that

$$dr=-\tfrac{1}{2}(GM)^{3/2}|\phi|^{-9/2}d\phi. \quad (536)$$

Hence,

$$C(\varrho)=\tfrac{1}{2}G^{3/2}\int_0^{\infty}dM\,M^{3/2}\tau(M)\int_{-\infty}^{+\infty}d\phi|\phi|^{-9/2}\times[1-\exp(i\varrho\cdot\phi)], \quad (537)$$

or, in an obvious notation

$$C(\varrho)=\tfrac{1}{2}G^{3/2}\langle M^{3/2}\rangle_{\text{Av}}\int_{-\infty}^{+\infty}[1-\exp(i\varrho\cdot\phi)]\times|\phi|^{-9/2}d\phi. \quad (538)$$

The foregoing expression is clearly unaffected if we replace ϕ by $-\phi$. But this replacement changes $\exp(i\varrho\cdot\phi)$ into $\exp(-i\varrho\cdot\phi)$ under the integral sign; taking the arithmetic mean of the two resulting integrals, we obtain

$$C(\varrho)=\tfrac{1}{2}G^{3/2}\langle M^{3/2}\rangle_{\text{Av}}\int_{-\infty}^{+\infty}[1-\cos(\varrho\cdot\phi)]|\phi|^{-9/2}d\phi. \quad (539)$$

Choosing polar coordinates with the z axis in the direction of ϱ Eq. (539) can be transformed to

$$C(\varrho)=\tfrac{1}{2}G^{3/2}\langle M^{3/2}\rangle_{\text{Av}}\int_0^{\infty}\int_{-1}^{+1}\int_0^{2\pi}\times[1-\cos(|\varrho||\phi|t)]|\phi|^{-5/2}d\omega dt d|\phi|, \quad (540)$$

or, introducing further the variable $z=|\varrho||\phi|$, we have

$$C(\varrho)=\tfrac{1}{2}G^{3/2}\langle M^{3/2}\rangle_{\text{Av}}|\varrho|^{3/2}\times\int_0^{\infty}\int_{-1}^{+1}\int_0^{2\pi}[1-\cos(zt)]z^{-5/2}d\omega dt dz. \quad (541)$$

After performing the integrations over ω and t we obtain

$$C(\varrho)=2\pi G^{3/2}\langle M^{3/2}\rangle_{\text{Av}}|\varrho|^{3/2}\times\int_0^{\infty}(z-\sin z)z^{-7/2}dz, \quad (542)$$

or after several integrations by parts

$$C(\varrho)=\frac{16}{15}\pi G^{3/2}\langle M^{3/2}\rangle_{\text{Av}}|\varrho|^{3/2}\int_0^{\infty}z^{-1/2}\cos z\,dz.$$
$$=\frac{4}{15}(2\pi G)^{3/2}\langle M^{3/2}\rangle_{\text{Av}}|\varrho|^{3/2}. \quad (543)$$

Combining Eqs. (529), (534), and (543) we now obtain

$$W(F)=\frac{1}{8\pi^3}\int_{-\infty}^{+\infty}\exp(-i\varrho\cdot F-a|\varrho|^{3/2})d\varrho, \quad (544)$$

where we have written

$$a=(4/15)(2\pi G)^{3/2}\langle M^{3/2}\rangle_{\text{Av}}n. \quad (545)$$

Using a frame of reference in which one of the principal axes is in the direction of F and changing to polar coordinates, the formula (544) for $W(F)$ can be reduced to

$$W(F)=\frac{1}{4\pi^2}\int_0^\infty\int_{-1}^{+1}\exp\left(-i|\varrho||F|t-a|\varrho|^{3/2}\right)\times|\varrho|^2dtd|\varrho|. \quad (546)$$

The integration over t is readily effected, and we obtain

$$W(F)=\frac{1}{2\pi^2|F|}\int_0^\infty\exp\left(-a|\varrho|^{3/2}\right)\times|\varrho|\sin\left(|\varrho||F|\right)d|\varrho|. \quad (547)$$

If we now put

$$x=|\varrho||F|, \quad (548)$$

Eq. (547) becomes

$$W(F)=\frac{1}{2\pi^2|F|^3}\int_0^\infty\exp\left(-ax^{3/2}/|F|^{3/2}\right)\times x\sin xdx. \quad (549)$$

We can rewrite the foregoing formula for $W(F)$ more conveniently if we introduce the *normal field* Q_H defined by

$$\left.\begin{aligned}Q_H&=a^{2/3}=(4/15)^{2/3}(2\pi G)(\langle M^{3/2}\rangle_{Av}n)^{2/3},\\&=2.6031G(\langle M^{3/2}\rangle_{Av}n)^{2/3}\end{aligned}\right\} \quad (550)$$

and express $|F|$ in terms of this unit:

$$|F|=\beta Q_H=\beta a^{2/3}. \quad (551)$$

Equation (549) now takes the form

$$W(F)=H(\beta)/4\pi a^2\beta^2, \quad (552)$$

where we have introduced the function $H(\beta)$ defined by

$$H(\beta)=\frac{2}{\pi\beta}\int_0^\infty\exp\left[-(x/\beta)^{3/2}\right]x\sin xdx. \quad (553)$$

Since,

$$W(|F|)=4\pi|F|^2W(F), \quad (554)$$

we obtain from Eqs. (551) and (552)

$$W(|F|)=H(\beta)/Q_H; \quad (555)$$

accordingly $H(\beta)$ defines the probability distribution of $|F|$ when it is expressed in units of Q_H. The function $H(\beta)$ has been evaluated numerically and is tabulated in Table IX.

The asymptotic behavior of the distribution $W(|F|)$ can be obtained from the formulae:

$$H(\beta)=4\beta^2/3\pi+O(\beta^4) \quad (\beta\rightarrow 0), \quad (556)$$

and

$$H(\beta)=(15/8)(2/\pi)^{1/2}\beta^{-5/2}+O(\beta^{-4}) \quad (\beta\rightarrow\infty). \quad (557)$$

We find [cf. Eqs. (551) and (555)]

$$W(|F|)\simeq(4/3\pi Q_H^3)|F|^2 \quad (|F|\rightarrow 0), \quad (558)$$

and

$$W(|F|)\simeq(15/8)(2/\pi)^{1/2}Q_H^{3/2}|F|^{-5/2} \quad (|F|\rightarrow\infty). \quad (559)$$

Substituting for Q_H from Eq. (550) in Eq. (559) we obtain

$$W(|F|)\simeq 2\pi G^{3/2}\langle M^{3/2}\rangle_{Av}n|F|^{-5/2} \quad (|F|\rightarrow\infty). \quad (560)$$

It is seen that while the frequency of occurrence of both the weak and the strong fields is quite small, it is only the fields of average intensity which have appreciable probabilities. In particular, the value of $|F|$ which has the maximum probability of occurrence is found to be (see Table IX) $\sim 1.6Q_H$.

Equations (552) and (553) provide, of course, the *exact* formula for the distribution of F for an *ideally* random distribution of stars. But an elementary treatment which leads to an approximate formula for $W(F)$ is of some interest and illuminates certain points in the theory. The treatment we refer to is based on the assumption that the force acting on a star is entirely due to its *nearest* neighbor.

Now, the law of distribution of the nearest neighbor is given by [see Appendix VII, Eq. (671)]

$$w(r)dr=\exp\left(-4\pi r^3n/3\right)4\pi r^2ndr, \quad (561)$$

and, since on the first neighbor approximation

$$|F|=GMr^{-2}, \quad (562)$$

we readily obtain the formula

$$W(|F|)d|F|=\exp\left[-4\pi(GM)^{3/2}n/3|F|^{3/2}\right]\times 2\pi(GM)^{3/2}n|F|^{-5/2}d|F|. \quad (563)$$

TABLE IX. The function $H(\beta)$.

β	$H(\beta)$	β	$H(\beta)$
0.0		5.0	0.04310
0.1	0.004225	5.2	0.03790
0.2	0.016666	5.4	0.03357
0.3	0.036643	5.6	0.02993
0.4	0.063084	5.8	0.02683
0.5	0.094601	6.0	0.02417
0.6	0.129598	6.2	0.02188
0.7	0.166380	6.4	0.01988
0.8	0.203270	6.6	0.01814
0.9	0.238704	6.8	0.01660
1.0	0.271322	7.0	0.01525
1.1	0.30003	7.2	0.01405
1.2	0.32402	7.4	0.01297
1.3	0.34281	7.6	0.01201
1.4	0.35620	7.8	0.01115
1.5	0.36426	8.0	0.01038
1.6	0.36726	8.2	0.00967
1.7	0.36566	8.4	0.00903
1.8	0.36004	8.6	0.00846
1.9	0.35101	8.8	0.00793
2.0	0.33918	9.0	0.00745
2.1	0.32519	9.2	0.00701
2.2	0.30951	9.4	0.00660
2.3	0.29266	9.6	0.00622
2.4	0.27485	9.8	0.00588
2.5	0.25667	10.0	0.00556
2.6	0.238	15.0	0.00188
2.7	0.222	20.0	0.00089
2.8	0.206	25.0	0.00050
2.9	0.190	30.0	0.00031
3.0	0.176	35.0	0.00021
3.2	0.150	40.0	0.00015
3.4	0.128	45.0	0.00011
3.6		50.0	0.00009
3.8		60.0	0.00005
4.0		70.0	0.00004
4.2		80.0	0.00003
4.4	0.06734	90.0	0.00002
4.6	0.05732	100.0	0.00002
4.8	0.04944		

According to the distribution (563)

$$W(|F|) \simeq 2\pi (GM)^{3/2} n |F|^{-5/2} \quad (|F| \to \infty), \quad (564)$$

which is seen to be in *exact* agreement with the formula (560) derived from the Holtsmark distribution (555). The physical meaning of this agreement, for $|F| \to \infty$ in the results derived from an exact and an approximate treatment of the same problem, is simply that the highest fields are in reality produced only by the nearest neighbor. More generally, it is found that the two distributions (555) and (563) agree over most of the range of $|F|$. Thus, the field which has the maximum frequency of occurrence on the basis of (563) is seen to differ from the corresponding value on the Holtsmark distribution by less than five percent. The region in which the two distributions (555) and (563) differ most markedly is when $|F| \to 0$: on the Holtsmark distribution $W(|F|)$ tends to zero as $|F|^2$ while on the nearest neighbor approximation $W(|F|)$ tends to zero as exp $(-\text{const.}\,|F|^{-3/2})$ [cf. Eqs. (558) and (564)]. However, the fact that the nearest neighbor approximation should be seriously in error for the weak fields is, of course, to be expected: for, a weak field arises from a more or less symmetrical, average, complexion of the stars around the one under consideration and consequently F under these circumstances is the result of the action of several stars and not due to any one single star.

Finally, we may draw attention to one important difficulty in using the Holstmark distribution for *all* values of $|F|$: It predicts relatively too high probabilities for $|F|$ as $|F| \to \infty$. Thus, on the basis of the distribution (555), $\langle |F|^2 \rangle_{\text{Av}}$ is divergent. [The same remark also applies to the distribution (564).] These relatively high probabilities for the high field strengths is a consequence of our assumption of complete randomness in stellar distribution for *all* elements of volume. It is, however, apparent that this assumption cannot be valid for the regions in the *very* immediate neighborhoods of the individual stars. For, if V denotes the relative velocity between two stars when separated by distances of the order of the average distance between the stars, the two stars cannot come closer together (on the approximation of linear trajectories) than a certain critical distance $r(|V|)$ such that

$$|V|^2/2 = [G(M_1 + M_2)/r(|V|)], \quad (565)$$

or

$$r(|V|) = [2G(M_1 + M_2)/|V|^2]. \quad (566)$$

Otherwise the two stars should be strictly regarded as the components of a binary system and this is inconsistent with our original premises. This restriction therefore leads us to infer that departures from true randomness exist for $r \sim r(|V|)$. However, under the conditions we normally encounter in stellar systems, $r(|V|)$ is very small compared to the average distance between the stars. Thus, in our galaxy, in the general neighborhood of the sun, $r(|V|) \sim 2 \times 10^{-5}$ parsec, and this is to be compared with an average distance between the stars of about three parsecs. Accordingly, we may expect the

Holtsmark distribution to be very close to the true distribution, except for the very highest values of $|F|$. More particularly, the deviations from the Holtsmark distribution are to be expected for field strengths of the order of

$$|F| \sim (GM_2/[r(|V|)]^2) \simeq (M_2[\langle |V|^2\rangle_{Av}]^2/4G(M_1+M_2)^2). \quad (567)$$

When $|F|$ becomes much larger than the quantity on the right-hand side of Eq. (567), the true frequencies of occurrence will very rapidly tend to zero as compared to what would be expected on the Holtsmark distribution, namely (560). A rigorous treatment of these deviations from the distribution (555) will require a reconsideration of the whole problem in *phase space* and is beyond the scope of the present investigation.

3. The Speed of Fluctuations in F

As we have already remarked the speed of fluctuations can be specified in terms of the distribution function $W(F, f)$ which gives the simultaneous probability of a given field strength F and an associated rate of change of F of amount f [cf. Eqs. (517) and (519)]. The general expression for this probability distribution can be readily written down using Markoff's method [I, §3, Eqs. (51), (52), and (53)]. We have [cf. Eqs. (529) and (530)]

$$W(F, f) = \frac{1}{64\pi^6}\int_{|\rho|=0}^{\infty}\int_{|\sigma|=0}^{\infty} \exp\left[-i(\rho\cdot F+\sigma\cdot f)\right]A(\rho, \sigma)d\rho d\sigma, \quad (568)$$

where

$$A(\rho, \sigma) = \underset{R\to\infty}{\text{limit}}\left[\frac{3}{4\pi R^3}\int_{0<M<\infty}\int_{|r|<R}\int_{|V|<\infty} \exp\left[i(\rho\cdot\phi+\sigma\cdot\psi)\right]\tau dr dV dM\right]^{4\pi R^3 n/3}. \quad (569)$$

In Eqs. (568) and (569) ρ and σ are two auxiliary vectors, n denotes the number of stars per unit volume, and

$$\phi = GM\frac{r}{|r|^3}; \quad \psi = GM\left\{\frac{V}{|r|^3} - 3\frac{r(r\cdot V)}{|r|^5}\right\}. \quad (570)$$

Further,

$$\tau dV dM = \tau(V; M)dV dM \quad (571)$$

gives the probability that a star with a relative velocity in the range $(V, V+dV)$ and with a mass between M and $M+dM$ will be found. It should also be noted that in writing down Eqs. (568) and (569) we have supposed (as in §2) that the fluctuations in the local stellar distribution which occur are subject only to the restriction of a constant average density.

Since

$$\frac{3}{4\pi R^3}\int_{M=0}^{\infty}\int_{|r|<R}\int_{|V|<\infty} \tau dr dV dM = 1, \quad (572)$$

we can rewrite (569) as

$$A(\rho, \sigma) = \underset{R\to\infty}{\text{limit}}\left\{1-\frac{3}{4\pi R^3}\int_{M=0}^{\infty}\int_{|r|<R}\int_{|V|<\infty}\{1-\exp\left[i(\rho\cdot\phi+\sigma\cdot\psi)\right]\}\tau dr dV dM\right\}^{4\pi R^3 n/3}. \quad (573)$$

The integral over r which occurs in Eq. (573) is seen to be conditionally convergent when extended over all $|r|$, i.e., also for $|r|\to\infty$. Hence, we can write

$$A(\rho, \sigma) = \underset{R\to\infty}{\text{limit}}\left\{1-\frac{3}{4\pi R^3}\int_{M=0}^{\infty}\int_{|r|=0}^{\infty}\int_{|V|=0}^{\infty}\{1-\exp\left[i(\rho\cdot\phi+\sigma\cdot\psi)\right]\}\tau dr dV dM\right\}^{4\pi R^3 n/3}, \quad (574)$$

or

$$A(\rho, \sigma) = \exp\left[-nC(\rho, \sigma)\right] \quad (575)$$

where

$$C(\varrho, \sigma)=\int_0^{\infty}\int_{-\infty}^{+\infty}\int_{-\infty}^{+\infty}\{1-\exp[i(\varrho\cdot\phi+\sigma\cdot\psi)]\}\tau d\tau dV dM. \tag{576}$$

This formally solves the problem. It does not, however, appear that the integral representing $C(\varrho, \sigma)$ can be evaluated explicitly in terms of any known functions. But if we are interested only in the moments of f for a given F and of F for a given f we need only the behavior of $A(\varrho, \sigma)$ and, therefore, also of $C(\varrho, \sigma)$ for $|\sigma|$, respectively, $|\varrho|$ tending to zero. For, considering the first and the second moments of the components f_ξ, f_η, and f_ζ of f along three directions ξ, η, and ζ at right angles to each other, we have

$$W(F)\langle f_\mu\rangle_{\text{Av}}=\int_{|f|=0}^{\infty}W(F, f)f_\mu df \quad (\mu=\xi, \eta, \zeta), \tag{577}$$

and

$$W(F)\langle f_\mu f_\nu\rangle_{\text{Av}}=\int_{|f|=0}^{\infty}W(F, f)f_\mu f_\nu df \quad (\mu, \nu=\xi, \eta, \zeta), \tag{578}$$

where $W(F)$ denotes the distribution of F for which we have already obtained an explicit formula in §2. Substituting now for $W(F, f)$ from Eq. (568) in the foregoing formulae for the moments we obtain

$$W(F)\langle f_\mu\rangle_{\text{Av}}=\frac{1}{64\pi^6}\int_{|f|=0}^{\infty}\int_{|\varrho|=0}^{\infty}\int_{|\sigma|=0}^{\infty}\exp[-i(\varrho\cdot F+\sigma\cdot f)]A(\varrho, \sigma)f_\mu d\varrho d\sigma df, \tag{579}$$

and

$$W(F)\langle f_\mu f_\nu\rangle_{\text{Av}}=\frac{1}{64\pi^6}\int_{|f|=0}^{\infty}\int_{|\varrho|=0}^{\infty}\int_{|\sigma|=0}^{\infty}\exp[-i(\varrho\cdot F+\sigma\cdot f)]A(\varrho, \sigma)f_\mu f_\nu d\varrho d\sigma df. \tag{580}$$

But

$$\left.\begin{aligned}
&\frac{1}{8\pi^3}\int_{|f|=0}^{\infty}\exp(-i\sigma\cdot f)f_\xi df=i\delta'(\sigma_\xi)\delta(\sigma_\eta)\delta(\sigma_\zeta),\\
&\frac{1}{8\pi^3}\int_{|f|=0}^{\infty}\exp(-i\sigma\cdot f)f_\xi^2 df=-\delta''(\sigma_\xi)\delta(\sigma_\eta)\delta(\sigma_\zeta),\\
&\frac{1}{8\pi^3}\int_{|f|=0}^{\infty}\exp(-i\sigma\cdot f)f_\xi f_\eta df=-\delta'(\sigma_\xi)\delta'(\sigma_\eta)\delta(\sigma_\zeta),
\end{aligned}\right\} \tag{581}$$

etc. In Eq. (581) δ denotes Dirac's δ-function and δ' and δ'' its first and second derivatives; and remembering also that

$$\int_{-\infty}^{+\infty}f(x)\delta(x)dx=f(0); \quad \int_{-\infty}^{+\infty}f(x)\delta'(x)dx=-f'(0); \quad \int_{-\infty}^{+\infty}f(x)\delta''(x)dx=f''(0), \tag{582}$$

Eqs. (579) and (580) for the moments reduce to

$$W(F)\langle f_\mu\rangle_{\text{Av}}=-\frac{i}{8\pi^3}\int_{|\varrho|=0}^{\infty}\exp(-i\varrho\cdot F)\left[\frac{\partial}{\partial\sigma_\mu}A(\varrho, \sigma)\right]_{|\sigma|=0}d\varrho, \tag{583}$$

and

$$W(F)\langle f_\mu f_\nu\rangle_{\text{Av}}=-\frac{1}{8\pi^3}\int_{|\varrho|=0}^{\infty}\exp(-i\varrho\cdot F)\left[\frac{\partial^2}{\partial\sigma_\mu\partial\sigma_\nu}A(\varrho, \sigma)\right]_{|\sigma|=0}d\varrho. \tag{584}$$

We accordingly see that the first and the second moments of f can be evaluated from a series expansion of $A(\varrho, \sigma)$ or of $C(\varrho, \sigma)$ which is correct up to the *second order* in $|\sigma|$. Such a series expansion has been found by Chandrasekhar and von Neumann and, quoting their final result, we have

$$C(\varrho, \sigma)=\frac{4}{15}(2\pi)^{\frac{3}{2}}G^{\frac{3}{2}}\langle M^{\frac{3}{2}}\rangle_{Av}|\varrho|^{\frac{3}{2}}+\frac{2}{3}\pi iG(\sigma_1\langle MV_1\rangle_{Av}+\sigma_2\langle MV_2\rangle_{Av}-2\sigma_3\langle MV_3\rangle_{Av})$$

$$+\frac{3}{28}(2\pi)^{\frac{1}{2}}G^{\frac{1}{2}}|\varrho|^{-\frac{1}{2}}[(5\sigma_1^2+4\sigma_2^2-2\sigma_3^2)\langle M^{\frac{1}{2}}V_1^2\rangle_{Av}+(4\sigma_1^2+5\sigma_2^2-2\sigma_3^2)\langle M^{\frac{1}{2}}V_2^2\rangle_{Av}$$

$$+(4\sigma_3^2-2\sigma_1^2-2\sigma_2^2)\langle M^{\frac{1}{2}}V_3^2\rangle_{Av}-8\sigma_2\sigma_3\langle M^{\frac{1}{2}}V_2V_3\rangle_{Av}-8\sigma_3\sigma_1\langle M^{\frac{1}{2}}V_3V_1\rangle_{Av}$$

$$+2\sigma_1\sigma_2\langle M^{\frac{1}{2}}V_1V_2\rangle_{Av}]+O(|\boldsymbol{\sigma}|^3) \quad (|\boldsymbol{\sigma}|\rightarrow 0), \quad (585)$$

where $\langle\ \rangle_{Av}$ indicates that the corresponding quantity has been averaged with the weight function $\tau(\boldsymbol{V}; M)$ [cf. Eq. (571)]; further, in Eq. (585) $(\sigma_1, \sigma_2, \sigma_3)$ and (V_1, V_2, V_3) are the components of $\boldsymbol{\sigma}$ and $\boldsymbol{V}$ in a system of coordinates in which the z axis is in the direction of $\boldsymbol{\varrho}$.

In Eq. (585) $\boldsymbol{V}=(V_1, V_2, V_3)$ is of course the velocity of a field star relative to the one under consideration. If we now let $\boldsymbol{u}$ and $\boldsymbol{v}$ denote the velocities of the field star and the star under consideration in an appropriately chosen local standard of rest, then

$$\boldsymbol{V}=\boldsymbol{u}-\boldsymbol{v}. \quad (586)$$

In their further discussion, Chandrasekhar and von Neumann introduce the assumption that the distribution of the velocities $\boldsymbol{u}$ among the stars is *spherical*, i.e., the distribution function $\Psi(\boldsymbol{u})$ has the form

$$\Psi(\boldsymbol{u})\equiv\Psi(j^2(M)|\boldsymbol{u}|^2), \quad (587)$$

where Ψ is an arbitrary function of the argument specified and the parameter j (of the dimensions of [velocity]$^{-1}$) can be a function of the mass of the star. This assumption for the distribution of the peculiar velocities $\boldsymbol{u}$ implies that the probability function $\tau(\boldsymbol{V}; M)$ must be expressible as

$$\tau(\boldsymbol{V}; M)\equiv\Psi[j^2(M)|\boldsymbol{u}|^2]\chi(M), \quad (588)$$

where $\chi(M)$ governs the distribution over the different masses. For a function τ of this form we clearly have

$$\langle MV_i\rangle_{Av}=-\langle M\rangle_{Av}v_i; \quad \langle M^{\frac{1}{2}}V_i^2\rangle_{Av}=\tfrac{1}{3}\langle M^{\frac{1}{2}}|\boldsymbol{u}|^2\rangle_{Av}+\langle M^{\frac{1}{2}}\rangle_{Av}v_i^2 \quad (i=1, 2, 3),$$
$$\langle M^{\frac{1}{2}}V_iV_j\rangle_{Av}=\langle M^{\frac{1}{2}}\rangle_{Av}v_iv_j \quad (i, j=1, 2, 3, i\neq j). \quad (589)$$

Substituting these values in Eq. (577) we find after some minor reductions that

$$C(\varrho, \boldsymbol{\sigma})=\frac{4}{15}(2\pi)^{\frac{3}{2}}G^{\frac{3}{2}}\langle M^{\frac{3}{2}}\rangle_{Av}|\varrho|^{\frac{3}{2}}-\frac{2}{3}\pi iG\langle M\rangle_{Av}(\sigma_1v_1+\sigma_2v_2-2\sigma_3v_3)+\frac{1}{4}(2\pi)^{\frac{1}{2}}G^{\frac{1}{2}}\langle M^{\frac{1}{2}}|\boldsymbol{u}|^2\rangle_{Av}|\varrho|^{-\frac{1}{2}}(\sigma_1^2+\sigma_2^2)$$

$$+\frac{3}{28}(2\pi)^{\frac{1}{2}}G^{\frac{1}{2}}\langle M^{\frac{1}{2}}\rangle_{Av}|\varrho|^{-\frac{1}{2}}[\sigma_1^2(5v_1^2+4v_2^2-2v_3^2)+\sigma_2^2(5v_2^2+4v_1^2-2v_3^2)$$

$$+\sigma_3^2(4v_3^2-2v_1^2-2v_2^2)+2\sigma_1\sigma_2v_1v_2-8\sigma_2\sigma_3v_2v_3-8\sigma_3\sigma_1v_3v_1]+O(|\boldsymbol{\sigma}|^3). \quad (590)$$

With a series expansion of this form we can, as we have already remarked, evaluate all the first and the second moments of f for a given $\boldsymbol{F}$.

Considering first the moment of f, Chandrasekhar and von Neumann find that

$$\langle f\rangle_{Av}=\overline{\left(\frac{dF}{dt}\right)}_{F,v}=-\frac{2}{3}\pi G\langle M\rangle_{Av}nB\left(\frac{|F|}{Q_H}\right)\left(v-3\frac{v\cdot F}{|F|^2}F\right), \quad (591)$$

where Q_H is the "normal field" introduced in §2 [Eqs. (550) and (551)] and

$$B(\beta)=3\left(\int_0^\beta H(\beta)d\beta \Big/ \beta H(\beta)\right)-1. \tag{592}$$

We shall examine certain formal consequences of Eq. (592).

Multiplying Eq. (591) scalarly with $\boldsymbol{F}$ we obtain

$$\boldsymbol{F}\cdot\overline{\left(\frac{d\boldsymbol{F}}{dt}\right)}_{\boldsymbol{F},\,\boldsymbol{v}}=\frac{4}{3}\pi G\langle M\rangle_{\text{Av}}nB\left(\frac{|\boldsymbol{F}|}{Q_H}\right)(\boldsymbol{v}\cdot\boldsymbol{F}); \tag{593}$$

but

$$\boldsymbol{F}\cdot\overline{\left(\frac{d\boldsymbol{F}}{dt}\right)}_{\boldsymbol{F},\,\boldsymbol{v}}=|\boldsymbol{F}|\overline{\left(\frac{d|\boldsymbol{F}|}{dt}\right)}_{\boldsymbol{F},\,\boldsymbol{v}}. \tag{594}$$

Hence,

$$\overline{\left(\frac{d|\boldsymbol{F}|}{dt}\right)}_{\boldsymbol{F},\,\boldsymbol{v}}=\frac{4}{3}\pi G\langle M\rangle_{\text{Av}}nB\left(\frac{|\boldsymbol{F}|}{Q_H}\right)\frac{\boldsymbol{v}\cdot\boldsymbol{F}}{|\boldsymbol{F}|}. \tag{595}$$

On the other hand, if F_j denotes the component of F in an arbitrary direction at right angles at the direction of $\boldsymbol{v}$ then according to Eq. (591)

$$\overline{\left(\frac{dF_j}{dt}\right)}_{\boldsymbol{F},\,\boldsymbol{v}}=2\pi G\langle M\rangle_{\text{Av}}nB\left(\frac{|\boldsymbol{F}|}{Q_H}\right)\frac{\boldsymbol{v}\cdot\boldsymbol{F}}{|\boldsymbol{F}|^2}F_j. \tag{596}$$

Combining Eqs. (595) and (596) we have

$$\frac{1}{F_j}\overline{\left(\frac{dF_j}{dt}\right)}_{\boldsymbol{F},\,\boldsymbol{v}}=\frac{3}{2}\frac{1}{|\boldsymbol{F}|}\overline{\left(\frac{d|\boldsymbol{F}|}{dt}\right)}_{\boldsymbol{F},\,\boldsymbol{v}}. \tag{597}$$

Equation (597) is clearly equivalent to

$$\overline{\frac{d}{dt}(\log F_j-\tfrac{3}{2}\log|\boldsymbol{F}|)_{\boldsymbol{F},\,\boldsymbol{v}}}=0. \tag{598}$$

We have thus proved that

$$\overline{\left[\frac{d}{dt}\left(\frac{F_j}{|\boldsymbol{F}|^{\frac{3}{2}}}\right)\right]}_{\boldsymbol{F},\,\boldsymbol{v}}=0. \tag{599}$$

We shall now examine the physical consequences of Eq. (591) more closely. In words, the meaning of this equation is that the component of

$$-\frac{2}{3}\pi G\langle M\rangle_{\text{Av}}nB\left(\frac{|\boldsymbol{F}|}{Q_H}\right)\left(\boldsymbol{v}-3\frac{\boldsymbol{v}\cdot\boldsymbol{F}}{|\boldsymbol{F}|^2}\boldsymbol{F}\right) \tag{600}$$

along any particular direction gives the average value of the rate of change of $\boldsymbol{F}$ that is to be expected in the specified direction when the star is moving with a velocity $\boldsymbol{v}$. Stated in this manner we at once see the essential difference in the stochastic variations of $\boldsymbol{F}$ with time in the two cases $|\boldsymbol{v}|=0$ and $|\boldsymbol{v}|\neq 0$. In the former case $\langle \boldsymbol{F}\rangle_{\text{Av}}\equiv 0$; but this is not generally true when $|\boldsymbol{v}|\neq 0$. Or expressed differently, when $|\boldsymbol{v}|=0$ the changes in $\boldsymbol{F}$ occur with equal probability in all directions while this is not the case when $|\boldsymbol{v}|\neq 0$. The true nature of this difference is brought out very clearly when we consider

$$\overline{\left(\frac{d|\boldsymbol{F}|}{dt}\right)}_{\boldsymbol{F},\,\boldsymbol{v}} \tag{601}$$

according to Eq. (595). Remembering that $B(\beta)\geqslant 0$ for $\beta\geqslant 0$, we conclude from Eq. (595) that

$$\overline{\left(\frac{d|\boldsymbol{F}|}{dt}\right)}_{\boldsymbol{F},\,\boldsymbol{v}}>0 \quad \text{if} \quad (\boldsymbol{v}\cdot\boldsymbol{F})>0, \tag{602}$$

and

$$\overline{\left(\frac{d|\boldsymbol{F}|}{dt}\right)}_{\boldsymbol{F},\,\boldsymbol{v}}<0 \quad \text{if} \quad (\boldsymbol{v}\cdot\boldsymbol{F})<0. \tag{603}$$

In other words, if $\boldsymbol{F}$ has a positive component in the direction of $\boldsymbol{v}$, $|\boldsymbol{F}|$ increases on the average, while if $\boldsymbol{F}$ has a negative component in the direction of $\boldsymbol{v}$, $|\boldsymbol{F}|$ decreases on the average. This essential asymmetry introduced by the direction of $\boldsymbol{v}$ may be expected to give rise to the phenomenon of *dynamical friction.*

Considering next the second moments of f Chandrasekhar and von Neumann find that

$$\langle|\boldsymbol{f}|^2{}_{\boldsymbol{F},\,\boldsymbol{v}}\rangle_{\text{Av}}=2ab\frac{\beta^{\frac{1}{2}}}{H(\beta)}\{2G(\beta)+7k[G(\beta)\sin^2\alpha-I(\beta)(3\sin^2\alpha-2)]\}$$
$$+\frac{g^2}{\beta H(\beta)}\{\beta H(\beta)(4-3\sin^2\alpha)+3K(\beta)(3\sin^2\alpha-2)\}, \tag{604}$$

where, α denotes the angle between the directions of $\boldsymbol{F}$ and $\boldsymbol{v}$

$$a=\frac{4}{15}(2\pi)^{\frac{3}{2}}G^{\frac{3}{2}}\langle M^{\frac{3}{2}}\rangle_{\text{Av}}n;\quad b=\frac{1}{4}(2\pi)^{\frac{3}{2}}G^{\frac{1}{2}}\langle M^{\frac{1}{2}}|\boldsymbol{u}|^2\rangle_{\text{Av}}n,\quad g=\frac{2}{3}\pi G\langle M\rangle_{\text{Av}}|\boldsymbol{v}|n;\quad k=\frac{3}{7}\frac{\langle M^{\frac{3}{2}}\rangle_{\text{Av}}|\boldsymbol{v}|^2}{\langle M^{\frac{1}{2}}|\boldsymbol{u}|^2\rangle_{\text{Av}}}, \tag{605}$$

and

$$\left.\begin{aligned} &H(\beta)=\frac{2}{\pi\beta}\int_0^\infty \exp\left[-(x/\beta)^{\frac{3}{2}}\right]\beta\sin\beta d\beta,\\ &G(\beta)=\frac{3}{2}\int_0^\beta \beta^{-\frac{1}{2}}H(\beta)d\beta,\qquad I(\beta)=\beta^{-\frac{3}{2}}\int_0^\beta \beta^{\frac{1}{2}}G(\beta)d\beta,\qquad K(\beta)=\int_0^\beta H(\beta)d\beta. \end{aligned}\right\} \tag{606}$$

Averaging Eq. (604) for all possible mutual orientations of the two vectors $\boldsymbol{F}$ and $\boldsymbol{v}$ we readily find that

$$\langle\langle|\boldsymbol{f}|^2{}_{\boldsymbol{F},\,|\boldsymbol{v}|}\rangle\rangle_{\text{Av}}=4ab\left\{\frac{\beta^{\frac{1}{2}}G(\beta)}{H(\beta)}\left(1+\frac{7}{3}k\right)+\frac{g^2}{2ab}\right\}, \tag{607}$$

or, substituting for k and $g^2/2ab$ from (605) we find

$$\langle\langle|\boldsymbol{f}|^2{}_{\boldsymbol{F}}\rangle\rangle_{\text{Av}}=4ab\left\{\frac{\beta^{\frac{1}{2}}G(\beta)}{H(\beta)}\left(1+\frac{\langle M^{\frac{3}{2}}\rangle_{\text{Av}}|\boldsymbol{v}|^2}{\langle M^{\frac{1}{2}}|\boldsymbol{u}|^2\rangle_{\text{Av}}}\right)+\frac{5}{12\pi}\frac{\langle M\rangle_{\text{Av}}{}^2|\boldsymbol{v}|^2}{\langle M^{\frac{3}{2}}\rangle_{\text{Av}}\langle M^{\frac{1}{2}}|\boldsymbol{u}|^2\rangle_{\text{Av}}}\right\}. \tag{608}$$

In terms of Eq. (608) we can define an approximate formula for the mean life of the state F according to the equation

$$T_{|\boldsymbol{F}|,\,|\boldsymbol{v}|}=|\boldsymbol{F}|/(\langle\langle|\boldsymbol{f}|^2{}_{\boldsymbol{F}}\rangle\rangle_{\text{Av}})^{\frac{1}{2}}. \tag{609}$$

Combining Eqs. (608) and (609) we find that

$$T_{|F|,\,|v|}=T_{|F|,\,0}\frac{1}{\left[1+\dfrac{\langle M^{\frac{1}{3}}\rangle_{\text{Av}}|v|^2}{\langle M^{\frac{1}{3}}|u|^2\rangle_{\text{Av}}}+\dfrac{5}{12\pi}\dfrac{\langle M\rangle_{\text{Av}}{}^2|v|^2}{\langle M^{\frac{1}{3}}\rangle_{\text{Av}}\langle M^{\frac{1}{3}}|u|^2\rangle_{\text{Av}}}\dfrac{H(\beta)}{\beta^{\frac{1}{3}}G(\beta)}\right]^{\frac{1}{2}}},\tag{610}$$

where $T_{|F|,\,0}$ denotes the mean life when $|v|=0$:

$$T_{|F|,\,0}=\left[\frac{a^{\frac{1}{3}}}{4b}\frac{\beta^{\frac{1}{3}}H(\beta)}{G(\beta)}\right]^{\frac{1}{2}}.\tag{611}$$

From Eq. (610) we derive that

$$T\propto|F|\quad\text{as}\quad|F|\rightarrow 0;\quad T\propto|F|^{-\frac{1}{2}}\quad\text{as}\quad|F|\rightarrow\infty;\tag{612}$$

in other words the mean life tends to zero for both weak and strong fields.

I am greatly indebted to Mrs. T. Belland for her assistance in preparing the manuscript for the press. My thanks are also due to Dr. L. R. Henrich for his careful revision of the entire manuscript.

APPENDIXES

I. THE MEAN AND THE MEAN SQUARE DEVIATION OF A BERNOULLI DISTRIBUTION

Consider the Bernoulli distribution

$$w(x)=\frac{n!}{x!(n-x)!}p^x(1-p)^{n-x}\quad(p<1;\ x\text{ a positive integer}\leqslant n).\tag{613}$$

An alternative form for $w(x)$ is

$$w(x)=C_x{}^n p^x q^{n-x},\tag{614}$$

where $C_x{}^n$ denotes the binomial coefficient and

$$q=1-p.\tag{615}$$

From Eq. (614) it is apparent that $w(x)$ is the coefficient of u^x in the expansion of $(pu+q)^n$:

$$w(x)=\text{coefficient of }u^x\text{ in }(pu+q)^n.\tag{616}$$

That $\sum w_x=1$ follows immediately from this remark:

$$\left.\begin{aligned}\sum_{x=1}^{n}w(x)&=\sum_{x=1}^{n}\text{coefficient of }u^x\text{ in }(pu+q)^n,\\&=[(pu+q)^n]_{u=1}=1.\end{aligned}\right\}\tag{617}$$

Consider now the mean and the mean square deviation of x. By definition

$$\langle x\rangle_{\text{Av}}=\sum_{x=1}^{n}xw(x)\tag{618}$$

and

$$\delta^2=\langle(x-\langle x\rangle_{\text{Av}})^2\rangle_{\text{Av}}=\langle x^2\rangle_{\text{Av}}-\langle x\rangle_{\text{Av}}{}^2=\sum_{x=1}^{n}x^2w(x)-\langle x\rangle_{\text{Av}}{}^2.\tag{619}$$

We have

$$\left.\begin{aligned}\langle x\rangle_{Av}&=\sum_{x=1}^{n}x\times\{\text{coefficient of } u^x \text{ in } (pu+q)^n\},\\ &=\sum_{x=1}^{n}\text{coefficient of } u^x \text{ in } \frac{d}{du}(pu+q)^n,\\ &=\left[\frac{d}{du}(pu+q)^n\right]_{u=1}=np(p+q).\end{aligned}\right\}\tag{620}$$

Hence

$$\langle x\rangle_{Av}=np.\tag{621}$$

Similarly,

$$\left.\begin{aligned}\langle x^2\rangle_{Av}&=\sum_{x=1}^{n}x^2\times\{\text{coefficient of } u^x \text{ in } (pu+q)^n\},\\ &=\sum_{x=1}^{n}\text{coefficient of } u^x \text{ in } \frac{d}{du}\left(u\frac{d}{du}[pu+q]^n\right),\\ &=\left\{\frac{d}{du}\left(u\frac{d}{du}[pu+q]^n\right)\right\}_{u=1},\end{aligned}\right\}\tag{622}$$

or,

$$\langle x^2\rangle_{Av}=np+n(n-1)p^2.\tag{623}$$

Combining Eqs. (619), (621) and (623) we obtain

$$\delta^2=np-np^2=np(1-p)=npq.\tag{624}$$

II. A PROBLEM IN PROBABILITY: MULTIVARIATE GAUSSIAN DISTRIBUTIONS

In Chapter I (§4, subsection [a]) we considered the special case of the problem of random flights in which the N displacements which the particle suffers are all governed by Gaussian distributions but with different variances. We shall now consider a generalization of this problem which has important applications to the theory of Brownian motion (see Chapter II, §2, lemma II).

Let

$$\mathbf{\Psi}=\sum_{j=1}^{N}\psi_j\mathbf{r};\quad \mathbf{\Phi}=\sum_{j=1}^{N}\phi_j\mathbf{r},\tag{625}$$

where the ψ_j's and the ϕ_j's are two arbitrary sets of N real numbers each, and where further $\mathbf{r}$ is a stochastic variable the probability distribution of which is governed by

$$\tau(\mathbf{r})=(1/(2\pi l^2)^{\frac{3}{2}})\exp(-|\mathbf{r}|^2/2l^2),\tag{626}$$

where l is a constant. We require the probability $W(\mathbf{\Psi},\mathbf{\Phi})d\mathbf{\Psi}d\mathbf{\Phi}$ that $\mathbf{\Psi}$ and $\mathbf{\Phi}$ shall lie, respectively, in the ranges $(\mathbf{\Psi},\mathbf{\Psi}+d\mathbf{\Psi})$ and $(\mathbf{\Phi},\mathbf{\Phi}+d\mathbf{\Phi})$. Applying Markoff's method to this problem, we have [cf. Eqs. (51) and (52)]

$$W(\mathbf{\Psi},\mathbf{\Phi})=\frac{1}{64\pi^6}\int_{-\infty}^{+\infty}\int_{-\infty}^{+\infty}\exp[-i(\boldsymbol{\varrho}\cdot\mathbf{\Psi}+\boldsymbol{\sigma}\cdot\mathbf{\Phi})]A_N(\boldsymbol{\varrho},\boldsymbol{\sigma})d\boldsymbol{\varrho}d\boldsymbol{\sigma},\tag{627}$$

where $\boldsymbol{\varrho}$ and $\boldsymbol{\sigma}$ are two auxiliary vectors and

$$A_N(\boldsymbol{\varrho},\boldsymbol{\sigma})=\prod_{j=1}^{N}\frac{1}{(2\pi l^2)^{\frac{3}{2}}}\int_{-\infty}^{+\infty}\exp[i(\boldsymbol{\varrho}\cdot\psi_j\mathbf{r}+\boldsymbol{\sigma}\cdot\phi_j\mathbf{r})]\exp(-|\mathbf{r}|^2/2l^2)d\mathbf{r}.\tag{628}$$

To evaluate $A_N(\boldsymbol{\varrho}, \boldsymbol{\sigma})$ we need the value of the typical integral

$$J = \frac{1}{(2\pi l^2)^{\frac{3}{2}}} \int_{-\infty}^{+\infty} \exp\left[i\mathbf{r}\cdot(\psi_j\boldsymbol{\varrho}+\phi_j\boldsymbol{\sigma}) - (|\mathbf{r}|^2/2l^2)\right] d\mathbf{r}. \tag{629}$$

We have

$$\begin{aligned} J &= \prod_{x,y,z} \frac{1}{(2\pi l^2)^{\frac{1}{2}}} \int_{-\infty}^{+\infty} \exp\left\{-\left[x^2+2il^2x(\rho_1\psi_j+\sigma_1\phi_j)\right]/2l^2\right\} dx, \\ &= \exp\left\{-l^2\left[(\rho_1\psi_j+\sigma_1\phi_j)^2+(\rho_2\psi_j+\sigma_2\phi_j)^2+(\rho_3\psi_j+\sigma_3\phi_j)^2\right]/2\right\}. \end{aligned} \tag{630}$$

Hence

$$\begin{aligned} A_N(\boldsymbol{\varrho}, \boldsymbol{\sigma}) &= \exp\left\{-l^2 \sum_{j=1}^{N} \left[(\rho_1\psi_j+\sigma_1\phi_j)^2+(\rho_2\psi_j+\sigma_2\phi_j)^2+(\rho_3\psi_j+\sigma_3\phi_j)^2\right]/2\right\} \\ &= \exp\left[-(P|\boldsymbol{\varrho}|^2+2R\boldsymbol{\varrho}\cdot\boldsymbol{\sigma}+Q|\boldsymbol{\sigma}|^2)/2\right], \end{aligned} \tag{631}$$

where we have written

$$P = l^2 \sum_{j=1}^{N} \psi_j^2; \quad R = l^2 \sum_{j=1}^{N} \phi_j\psi_j; \quad Q = l^2 \sum_{j=1}^{N} \phi_j^2. \tag{632}$$

Substituting for $A_N(\boldsymbol{\varrho}, \boldsymbol{\sigma})$ from Eq. (632) in the formula for $W(\boldsymbol{\Psi}, \boldsymbol{\Phi})$ [Eq. (627)] we obtain

$$W(\boldsymbol{\Psi}, \boldsymbol{\Phi}) = \frac{1}{64\pi^6} \prod_{i=1}^{3} \int_{-\infty}^{+\infty} \int_{-\infty}^{+\infty} \exp\left\{-\left[P\rho_i^2+2R\rho_i\sigma_i+Q\sigma_i^2+2i(\rho_i\Psi_i+\sigma_i\Phi_i)\right]/2\right\} d\rho_i d\sigma_i. \tag{633}$$

To evaluate the integrals occurring in the foregoing formula, we first perform a translation of the coordinate system according to

$$\rho_i = \xi_i + \alpha_i; \quad \sigma_i = \eta_i + \beta_i \quad (i = 1, 2, 3), \tag{634}$$

where α_i and β_i are so chosen that

$$P\alpha_i + R\beta_i = -i\Psi_i; \quad R\alpha_i + Q\beta_i = -i\Phi_i \quad (i = 1, 2, 3). \tag{635}$$

With this transformation of the variables we have

$$\begin{aligned} P\rho_i^2+2R\rho_i\sigma_i+Q\sigma_i^2+2i(\rho_i\Psi_i+\sigma_i\Phi_i) &= P\xi_i^2+2R\xi_i\eta_i+Q\eta_i^2+i(\alpha_i\Psi_i+\beta_i\Phi_i), \\ &= P\xi_i^2+2R\xi_i\eta_i+Q\eta_i^2+\frac{1}{PQ-R^2}(P\Phi_i^2-2R\Phi_i\Psi_i+Q\Psi_i^2). \end{aligned} \tag{636}$$

Hence,

$$\begin{aligned} W(\boldsymbol{\Psi}, \boldsymbol{\Phi}) = \frac{1}{64\pi^6} \prod_{i=1}^{3} &\exp\left[-(P\Phi_i^2-2R\Phi_i\Psi_i+Q\Psi_i^2)/2(PQ-R^2)\right] \\ &\times \int_{-\infty}^{+\infty} \int_{-\infty}^{+\infty} \exp\left[-(P\xi_i^2+2R\xi_i\eta_i+Q\eta_i^2)/2\right] d\xi_i d\eta_i. \end{aligned} \tag{637}$$

From this equation we readily find that

$$W(\boldsymbol{\Psi}, \boldsymbol{\Phi}) = \left[1/8\pi^3(PQ-R^2)^{\frac{3}{2}}\right] \exp\left[-(P|\boldsymbol{\Phi}|^2-2R\boldsymbol{\Psi}\cdot\boldsymbol{\Phi}+Q|\boldsymbol{\Psi}|^2)/2(PQ-R^2)\right], \tag{638}$$

which gives the required probability distribution.

III. THE POISSON DISTRIBUTION AS THE LAW OF DENSITY FLUCTUATIONS

Consider an element of volume v which is a part of a larger volume V. Let there be N particles distributed in a random fashion inside the volume V. Under these conditions the probability that a particular particle will be found in the element of volume v is clearly v/V; similarly, the probability that it will *not* be found inside v is $(V-v)/V$. Hence, the probability $W_N(n)$ that *some n* particles will be found inside v is given by the Bernoulli distribution

$$W_N(n)=\frac{N!}{n!(N-n)!}\left(\frac{v}{V}\right)^n\left(1-\frac{v}{V}\right)^{N-n}. \tag{639}$$

The average value of n is therefore given by [cf. Eq. (621)]

$$\langle n\rangle_{\text{Av}}=N(v/V)=\nu \quad \text{(say)}. \tag{640}$$

In terms of ν Eq. (639) can be expressed in the form

$$W_N(n)=\frac{N!}{n!(N-n)!}\left(\frac{\nu}{N}\right)^n\left(1-\frac{\nu}{N}\right)^{N-n}. \tag{641}$$

The case of greatest practical interest arises when both N and V tend to infinity but in such a way that ν remains constant [see Eq. (640)]. To obtain the corresponding limiting form of the distribution (641) we first rewrite it as

$$\left.\begin{aligned} W_N(n)&=\frac{1}{n!}N(N-1)(N-2)\cdots(N-n+1)\left(\frac{\nu}{N}\right)^n\left(1-\frac{\nu}{N}\right)^{N-n},\\ &=\frac{\nu^n}{n!}1\left(1-\frac{1}{N}\right)\left(1-\frac{2}{N}\right)\cdots\left(1-\frac{n-1}{N}\right)\left(1-\frac{\nu}{N}\right)^{N-n}, \end{aligned}\right\} \tag{642}$$

and then let $N\rightarrow\infty$ keeping both ν and n fixed. We have

$$\left.\begin{aligned} W(n)&=\underset{N\rightarrow\infty}{\text{limit}}\ W_N(n),\\ &=\frac{\nu^n}{n!}\underset{N\rightarrow\infty}{\text{limit}}\left\{\left(1-\frac{1}{N}\right)\left(1-\frac{2}{N}\right)\cdots\left(1-\frac{n-1}{N}\right)\left(1-\frac{\nu}{N}\right)^{N-n}\right\},\\ &=\frac{\nu^n}{n!}\underset{N\rightarrow\infty}{\text{limit}}\left(1-\frac{\nu}{N}\right)^N. \end{aligned}\right\} \tag{643}$$

Hence,

$$W(n)=\nu^n e^{-\nu}/n!, \tag{644}$$

which is the required Poisson distribution.

In some applications of Eq. (644) (e.g., III, §3) ν is a very large number; and when this is the case, interest is attached to only those values of n which are relatively close to ν. We shall now show that under these conditions the Poisson distribution specializes still further to a Gaussian distribution.

Rewriting Eq. (644) in the form

$$\log W(n)=n\log\nu-\nu-\log n! \tag{645}$$

and adopting Stirling's approximation for $\log n$ [cf. Eq. (7)] we obtain

$$\log W(n)=n\log\nu-\nu-(n+\tfrac{1}{2})\log n+n-\tfrac{1}{2}\log 2\pi+O(n^{-1}). \tag{646}$$

Let

$$n=\nu+\delta. \tag{647}$$

Equation (646) becomes

$$\log W(n)=-(\nu+\delta+\tfrac{1}{2})\log\left(1+\frac{\delta}{\nu}\right)+\delta-\tfrac{1}{2}\log(2\pi\nu)+O(n^{-1}). \tag{648}$$

If we now suppose that $\delta/\nu \ll 1$ we can expand the logarithmic term in Eq. (648) as a power series in δ/ν. Retaining only the dominant term, we find

$$\log W(n) = -(\delta^2/2\nu) - \tfrac{1}{2}\log(2\pi\nu) \quad (\nu\to\infty;\ \delta/\nu\to 0). \tag{649}$$

Thus,

$$W(n) = [1/(2\pi\nu)^{\frac{1}{2}}]\exp[-(n-\nu)^2/2\nu], \tag{650}$$

which is the required Gaussian form.

IV. THE MEAN AND THE MEAN SQUARE DEVIATION OF THE SUM OF TWO PROBABILITY DISTRIBUTIONS

Let $w_1(x)$ and $w_2(y)$ represent two probability distributions. For the sake of definiteness we shall suppose that x and y take on only discrete values. A probability distribution which is said to be the *sum* of the two distributions is defined by

$$w(z) = \sum_{x+y=z} w_1(x)w_2(y), \tag{651}$$

where in the summation on the right-hand side we include all pairs of values of x and y (each in their respective domains) which satisfy the relation $x+y=z$. We may first verify that $w(z)$ defined according to Eq. (651) does in fact represent a probability distribution. To see this we have only to show that $\sum w(z)=1$. Now,

$$\sum_z w(z) = \sum_z \sum_{x+y=z} w_1(x)w_2(y); \tag{652}$$

accordingly, in the summation on the right-hand side, x and y can now run through their respective ranges of values *independently* of each other. Hence,

$$\sum_z w(z) = [\sum_x w_1(x)][\sum_y w_2(y)] = 1. \tag{653}$$

We shall now prove that *the mean and the mean square deviation of the sum of two probability distributions is the sum of the means and the mean square deviations of the component distributions.*

To prove this theorem, we observe that by definitions

$$\langle z\rangle_{\text{Av}} = \sum_z zw(z) = \sum_z \sum_{x+y=z} (x+y)w_1(x)w_2(y), \tag{654}$$

or

$$\langle z\rangle_{\text{Av}} = \sum_x \sum_y [xw_1(x)w_2(y) + yw_1(x)w_2(y)], \tag{655}$$

where in the summations on the right-hand side we can again let x and y run their respective ranges of values independently of each other. Hence,

$$\langle z\rangle_{\text{Av}} = [\sum_x xw_1(x)][\sum_y w_2(y)] + [\sum_x w_1(x)][\sum_y yw_2(y)], \tag{656}$$

or

$$\langle z\rangle_{\text{Av}} = \langle x\rangle_{\text{Av}} + \langle y\rangle_{\text{Av}}. \tag{657}$$

Similarly,

$$\left.\begin{aligned}
\langle (z-\langle z\rangle_{\text{Av}})^2\rangle_{\text{Av}} &= \sum_z (z-\langle z\rangle_{\text{Av}})^2 w(z),\\
&= \sum_z \sum_{x+y=z} (x+y-\langle x\rangle_{\text{Av}}-\langle y\rangle_{\text{Av}})^2 w_1(x)w_2(y).\\
&= \sum_x \sum_y [(x-\langle x\rangle_{\text{Av}})^2 + 2(x-\langle x\rangle_{\text{Av}})(y-\langle y\rangle_{\text{Av}}) + (y-\langle y\rangle_{\text{Av}})^2]w_1(x)w_2(y),\\
&= [\sum_x (x-\langle x\rangle_{\text{Av}})^2 w_1(x)][\sum_y w_2(y)] + [\sum_x w_1(x)][\sum_y (y-\langle y\rangle_{\text{Av}})^2 w_2(y)]\\
&\quad + 2[\sum_x (x-\langle x\rangle_{\text{Av}})w_1(x)][\sum_y (y-\langle y\rangle_{\text{Av}})w_2(y)].
\end{aligned}\right\} \tag{658}$$

Hence,

$$\langle (z-\langle z\rangle_{Av})^2\rangle_{Av}=\langle (x-\langle x\rangle_{Av})^2\rangle_{Av}+\langle (y-\langle y\rangle_{Av})^2\rangle_{Av}. \tag{659}$$

The theorem is now proved.

The extension of the foregoing results to include the case when x and y are continuously variable is, of course, obvious. Similarly the definitions and results can be further extended to include the sums of more than two probability distributions.

V. ZERMELO'S PROOF OF POINCARÉ'S THEOREM CONCERNING THE QUASI-PERIODIC CHARACTER OF THE MOTIONS OF A CONSERVATIVE DYNAMICAL SYSTEM

Consider a conservative dynamical system of n degrees of freedom and which is described by a Hamiltonian function H of the generalized coordinates $q_1, \cdots, q_n$ and momenta $p_1, \cdots, p_n$. The state of such a dynamical system can be represented by a point in the $2n$ dimensional phase space of the q's and p's. Similarly, the trajectory described by the representative point will describe the evolution of the dynamical system.

Through each point in the phase space there passes a unique trajectory which can be derived from the canonical equations of motion

$$\dot{q}_s=\frac{\partial H}{\partial p_s};\quad \dot{p}_s=-\frac{\partial H}{\partial q_s}\quad (s=1,\cdots,n). \tag{660}$$

More generally, consider any arbitrary continuous domain of points g_0 (of finite measure) in the phase space. Let the points g_0 be the representatives at time $t=0$ of an ensemble of dynamical systems all described by the same Hamiltonian function $H(p_1, \cdots, p_n; q_1, \cdots, q_n)$. At a later time t the representatives of the ensemble will occupy a continuous domain of points g_t which can be obtained by tracing through each point of g_0 the corresponding trajectory and following along the various trajectories for a time t. Because of the uniqueness, in general, of the trajectories passing through a given point in the phase space, the construction of the domain g_t from an initial domain g_0 is a unique process. We shall accordingly refer to g_t as the *future phase* (at time t) of the *initial phase* g_0 (at time $t=0$) of the given dynamical system.

Now, according to Liouville's theorem of classical dynamics, the density of any element of phase space remains constant during its motion according to the canonical Eqs. (660). Hence, if ω_t denotes the volume extension of the domain of points g_t introduced in the preceding paragraph, it follows from Liouville's theorem that ω_t remains constant as t varies.

We have already described how from an initial phase g_0 we can derive the future phase g_t at time t. The domain of points g_0 together with *all* its future phases g_t, $(0<t<\infty)$ clearly form a continuous domain of points which we shall denote by Γ_0: Γ_0 is accordingly the class of all states which at some finite past occupied states belonging to g_0. The extension of Γ_0 will be finite if we are considering a dynamical system which is enclosed—for, then, none of the coordinates or momenta can take on infinite values and the entire accessible region of the phase space remains finite. We shall suppose that this is the case and denote by Ω_0 the extension of Γ_0. Clearly $\Omega_0 \geqslant \omega_0$. In a similar manner we can, quite generally, define the domain of points Γ_t which includes all the future phases of g_t. Let Ω_t denote the extension of Γ_t. It is evident that

$$\Omega_{t_1} \geqslant \Omega_{t_2} \quad \text{whenever} \quad t_1<t_2. \tag{661}$$

For, Ω_{t_1} denoting the extension of *all* the future phases of g_{t_1} must therefore necessarily include also the future phases of g_{t_2} if $t_1<t_2$. On the other hand, considering Γ_0 itself as a domain of points, we can construct *its* future phases in exactly the same way as the future phases g_t of g_0 were constructed. But the future phase of Γ_0 after a time t is clearly Γ_t. And therefore applying Liouville's theorem to the domain Γ_0 and its future phases Γ_t we conclude that

$$\Omega_t=\text{constant}. \tag{662}$$

Comparing this result with the inequality (661) we infer that *the domain of points Γ_t can differ from Γ_0 by at most a set of points of measure zero.* Hence, the future phases of $g_t (t>0$ but arbitrary otherwise) must include g_0 apart, possibly, from a set of points of measure zero. But the points of g_t are themselves future phases of the points of g_0. Hence, the states belonging to g_0 (again, with the possible exception of a set of zero measure) must recur after the elapse of a sufficient length of time; and this is true no matter how small the extension ω_0 of g_0 is, provided it is only finite. From this, the deduction of Poincaré's theorem is immediate. (For a formal statement of Poincaré's theorem see Chapter III, §4).

VI. BOLTZMANN'S ESTIMATE OF THE PERIOD OF A POINCARÉ CYCLE

To estimate the order of magnitude of the period of a Poincaré cycle, Boltzmann has considered the following typical example:

A cubic centimeter of air containing 10^{18} molecules is considered in which all the molecules are initially supposed to have a speed of 500 meters per second. With a concentration of 10^{18} molecules, the average distance between the neighboring ones is of the order of 10^{-6} cm. Also, under normal conditions, each molecule will suffer something like 4×10^9 collisions per second so that on the whole there will occur

$$b=2\times10^{27} \text{ collisions per second.} \tag{663}$$

Since Poincaré's theorem asserts only the quasi-periodic character of the motions (see Chapter III, §4 and Appendix V) the period to be estimated clearly depends on the closeness to which we require the initial conditions to recur. For the case under discussion Boltzmann supposes that a molecule can be said to have approximately returned to its initial state if the differences in position (x, y, z) and velocity (u, v, w) in the initial and the final states are such that

$$|\Delta x|,\ |\Delta y|,\ |\Delta z| \leqslant 10^{-7} \text{ cm}, \tag{664}$$

and

$$|\Delta u|,\ |\Delta v|,\ |\Delta w| \leqslant 1 \text{ m/sec.} \tag{665}$$

In other words, we shall require the positions to agree to within 10 percent of the average distance between the molecules and the velocities to agree within one part in 500.

We shall first estimate the order of magnitude of the time required for the recurrence of an initial "abnormal" distribution in the velocities. According to Poincaré's theorem, an initial state need not recur earlier than the time necessary for all the molecules to take on all the possible values for the velocity. We can readily determine the number N of such possibilities with the understanding that we agree to distinguish between two velocities only if at least one of the components differ by more than 1 m/sec.

The first molecule can have all velocities ranging from zero to $a=500\times10^9$ m/sec.—since we have supposed that in the initial state all the molecules have the same speed of 500 m/sec and that there are 10^{18} molecules in the system. Again, if the first molecule has a speed v_1 the second one can have speeds only in range 0 to $(a^2-v_1^2)^{\frac{1}{2}}$. Similarly, if the first and the second molecules have speeds v_1 and v_2, respectively, the third molecule can have speeds only in the range 0 to $(a^2-v_1^2-v_2^2)^{\frac{1}{2}}$; and so on. Accordingly, the required number of combinations N is

$$\left.\begin{aligned} N &= (4\pi)^{n-1}\int_0^a dv_1 v_1^2 \int_0^{(a^2-v_1^2)^{\frac{1}{2}}} dv_2 v_2^2 \int_0^{(a^2-v_1^2-v_2^2)^{\frac{1}{2}}} dv_3 v_3^2 \cdots \int_0^{(a^2-v_1^2\cdots-v^2_{n-2})^{\frac{1}{2}}} dv_{n-1} v^2_{n-1}, \\ &= (\pi^{(3n-3)/2}/2\cdot3\cdot4\cdots[3(n-1)/2])a^{3(n-1)} \quad (n,\text{ odd}), \\ &= (2(2\pi)^{(3n-4)/2}/3\cdot5\cdot7\cdots3(n-1))a^{3(n-1)} \quad (n,\text{ even}), \end{aligned}\right\} \tag{666}$$

where

$$a=500\times10^9 \quad \text{and} \quad n=10^{18}. \tag{667}$$

Since each of these N combinations occurs on the average in a time $1/b$ seconds [cf. Eq. (663)] the total time required for the velocities to run through all the possible values is

$$N/b. \tag{668}$$

After this length of time we may expect the initial distribution of the velocities to recur to within the limits of accuracy specified except for one single molecule the direction of whose motion has been left unrestricted. On the other hand we have still left unspecified the positions of the centers of gravity of all the molecules. But in order that we may say that the initial state has recurred to a sufficient approximation, we must require the positions of the molecules in the final state also to agree with the initial values to some stated degree of accuracy. This would clearly require the time (668) to be multiplied by another number of order similar to N. However, the extremely large value already of N/b gives some indication of the enormous times which are involved. Moreover, comparing these times with the time of relaxation of a gas which is of the order 10^{-8} second under normal conditions, we get an idea as to how extremely small the fraction of the total number of complexions is for which appreciable departures from a Maxwellian distribution occur. (For a further discussion of these and related matters see Chapter III, §4.)

VII. THE LAW OF DISTRIBUTION OF THE NEAREST NEIGHBOR IN A RANDOM DISTRIBUTION OF PARTICLES

This problem was first considered by Hertz (see reference 71 in the Bibliographical Notes for Chapter IV).

Let $w(r)dr$ denote the probability that the nearest neighbor to a particle occurs between r and $r+dr$. This probability must be clearly equal to the probability that no particles exist interior to r times the probability that a particle does exist in the spherical shell between r and $r+dr$. Accordingly, the function $w(r)$ must satisfy the relation

$$w(r)=\left[1-\int_0^r w(r)dr\right]4\pi r^2 n, \tag{669}$$

where n denotes the average number of particles per unit volume. From Eq. (669) we derive:

$$\frac{d}{dr}\left[\frac{w(r)}{4\pi r^2 n}\right]=-4\pi r^2 n\frac{w(r)}{4\pi r^2 n}. \tag{670}$$

Hence

$$w(r)=\exp\left(-4\pi r^3 n/3\right)4\pi r^2 n, \tag{671}$$

since, according to Eq. (669)

$$w(r)\rightarrow 4\pi r^2 n \quad \text{as} \quad r\rightarrow 0. \tag{672}$$

Equation (671) gives then the required law of distribution of the nearest neighbor.

Using the distribution (671) we can derive an *exact* formula for the "average distance" D between the particles. For, by definition

$$D=\int_0^\infty rw(r)dr, \tag{673}$$

or, if we use Eq. (671)

$$D=\int_0^\infty \exp\left(-4\pi r^3 n/3\right)4\pi r^3 n dr. \tag{674}$$

After some elementary reductions, Eq. (674) becomes

$$\left.\begin{aligned} D &= \frac{1}{(4\pi n/3)^{\frac{1}{3}}}\int_0^\infty e^{-x}x^{\frac{1}{3}}dx,\\ &= \Gamma(4/3)/(4\pi n/3)^{\frac{1}{3}}.\end{aligned}\right\} \quad (675)$$

Substituting for $\Gamma(4/3)$, we find

$$D = 0.55396n^{-\frac{1}{3}}. \quad (676)$$

BIBLIOGRAPHICAL NOTES

Chapter I

§*1*.—We may briefly record here the history of the problem of random flights considered in this chapter:

Karl Pearson appears to have been the first to explicitly formulate a problem of this general type:

1. K. Pearson, Nature **77**, 294 (1905). Pearson's formulation of the problem was in the following terms: "A man starts from a point O and walks l yards in a straight line; he then turns through any angle whatever and walks another l yards in a second straight line. He repeats this process n times. I require the probability that after these n stretches he is at a distance between r and $r+dr$ from his starting point O." After Pearson had formulated this problem Lord Rayleigh pointed out that the problem is formally "the same as that of the composition of n isoperiodic vibrations of unit amplitude and of phases distributed at random" which he had considered as early as in 1880:

2. Lord Rayleigh, Phil. Mag. **10**, 73 (1880); see also ibid. **47**, 246 (1889). These papers are reprinted in *Scientific Papers of Lord Rayleigh*, Vol. I, p. 491, and Vol. IV, p. 370. In the foregoing papers Rayleigh obtains the asymptotic form of the solution as $n\rightarrow\infty$. But for finite values of n the general solution of Pearson's problem was given by

3. J. C. Kluyver, Konink. Akad. Wetenschap. Amsterdam **14**, 325 (1905). The general solution of the problem of random walk in one dimension was obtained by Smoluchowski apparently independently of the earlier investigators.

4. M. von Smoluchowski, Bull. Acad. Cracovie, p. 203 (1906). In its most general form the problem of random flights was formulated by A. A. Markoff who also outlined the method for obtaining the general solution.

5. A. A. Markoff, *Wahrscheinlichkeitsrechnung* (Leipzig, 1912), §§16 and 33.

§*2*.—The problem of the random walk with reflecting and absorbing barriers was first considered by Smoluchowski:

6. M. v. Smoluchowski, (a) Wien Ber. **124**, 263 (1915); also (b) "Drei Vortrage uber Diffusion, Brownsche Bewegung und Koagulation von Kolloidteilchen," Physik. Zeits. **17**, 557, 585 (1916). See also

7. R. von Mises, *Wahrscheinlichkeitsrechnung* (Leipzig and Wien), pp. 479–518.

§*3*.—Markoff's method described in this section is a somewhat generalized version of what is given in Markoff (reference 5). See also

8. M. von Laue, Ann. d. Physik **47**, 853 (1915).

§*4*.—See A. A. Markoff (reference 5). The case of finite N considered in subsection (b) follows the treatment of

9. Lord Rayleigh, Phil. Mag. **37**, 321 (1919) (or *Scientific Papers*, Vol. VI, p. 604).

§*5*.—The passage to a differential equation for the case of the one-dimensional problem of the random walk was achieved by Rayleigh:

10. Lord Rayleigh, Phil. Mag. **47**, 246 (1899) (or *Scientific Papers*, Vol. IV, p. 370). See also Smoluchowski (reference 6). But the general treatment given in this section appears to be new.

We may also note the following further reference:

11. W. H. McCrea, Proc. Roy. Soc. Edinburgh **60**, 281 (1939).

Chapter II

The following general references may be noted.

12. The Svedberg, *Die Existenz der Molekule* (Leipzig, 1912).

13. G. L. de Haas-Lorentz, *Die Brownsche-Bewegung und einige verwandte Erscheinungen*, (Braunschweig, 1913).

14. M. v. Smoluchowski, see reference 6(b).

15. J. Perrin, *Atoms* (Constable, London, 1916).

16. R. Fürth, *Schwankungserscheinungen in der Physik* (Sammlung Vieweg, Braunschweig, 1920), Vol. 48.

§*1*.—As is well known the modern theory of Brownian motion was initiated by Einstein and Smoluchowski:

17. A. Einstein, Ann. d. Physik **17**, 549 (1905); also, ibid. **19**, 371 (1906).

18. M. v. Smoluchowski, Ann. d. Physik **21**, 756 (1906). In Einstein's and in Smoluchowski's treatment of the problem, Brownian motion is idealized as a problem in random flights; but as we have seen, this idealization is valid only when we ignore effects which occur in time intervals of order β^{-1}. For the general treatment of the problem we require to base our discussion on an equation of the type first introduced by Langevin:

19. P. Langevin, Comptes rendus **146**, 530 (1908). In this connection see

20. F. Zernike, *Handbuch der Physik* (Berlin, 1928), Vol. 3, p. 456.

§*2*.—The treatment of the Brownian motion of a free particle given in this section is derived from:

21. L. S. Ornstein and W. R. van Wijk, Physica **1**, 235 (1933). See also

22. W. R. van Wijk, Physica **3**, 1111 (1936). Earlier, but somewhat less general treatment along the same lines is contained in

23. G. E. Uhlenbeck and L. S. Ornstein, Phys. Rev. **36**, 823 (1930). In the foregoing papers the discussion has been carried out only for the case of one-dimensional motion. In the text we have treated the general three-dimensional problem; further, the arguments in references 21 and 22 have been rearranged considerably to make the presentation more direct and straightforward.

§*3*.—See Ornstein and Wijk (reference 21); also

24. G. E. Uhlenbeck and S. Goudsmidt, Phys. Rev. **34**, 145 (1929).

25. G. A. van Lear and G. E. Uhlenbeck, Phys. Rev. **38**, 1583 (1931).

§*4*.—The passage to a differential equation for the description of the Brownian motion of a free particle in the velocity space was achieved by

26. A. D. Fokker, Ann. d. Physik **43**, 812 (1914). A more general discussion of this problem is due to

27. M. Planck, Sitz. der preuss. Akad. p. 324 (1917). See also references 21 and 23; further,

28. Lord Rayleigh, *Scientific Papers*, Vol. III, p. 473.

29. L. S. Ornstein, Versl. Acad. Amst. **26**, 1005 (1917); also Konink. Akad. Wetenschap. Amsterdam **20**, 96 (1917).

30. H. C. Burger, Versl. Acad. Amst. **25**, 1482 (1917).

31. L. S. Ornstein and H. C. Burger, Versl. Acad. Amst. **27**, 1146 (1919); **28**, 183 (1919); also Konink. Akad. Wetenschap. Amsterdam **21**, 922 (1918).

Earlier attempts to generalize Liouville's equation of classical dynamics to include Brownian motion are contained in

32. O. Klein, Arkiv for Matematik, Astronomi, och Fysik **16**, No. 5 (1921); and

33. H. A. Kramers, Physica **7**, 284 (1940).

The passage to a differential equation in configuration space was first achieved by

34. M. v. Smoluchowski, Ann. d. Physik **48**, 1103 (1915); see also,

35. R. Fürth, Ann. d. Physik **53**, 177 (1917).

In the text the discussion of the various differential equations has been carried out more generally and more completely than in the references given above; this is particularly true of the discussion relating to the generalization of the Liouville equation of classical dynamics (subsections, ii–v).

§*5*.—See H. A. Kramers (reference 33).

Approaches to the problem of the Brownian motion somewhat different to the one we have adopted are contained in

36. G. Krutkov, Physik. Zeits. der Sowjetunion **5**, 287 (1934). See also the various articles by the same author in C. R. Acad. Sci. USSR during the years (1934) and (1935).

37. S. Bernstein, C. R. Acad. Sci. USSR, p. 1 (1934), and p. 361 (1934). A more particularly mathematical discussion of the problems of Brownian motion has been given by

38. J. L. Doob, Ann. Math. **43**, 351 (1942); see also the references given in this paper.

Chapter III

The following general references may be noted.

39. M. v. Smoluchowski, reference 6(b).

40. A. Sommerfeld, "Zum Andenken an Marian von Smoluchowski," Physik. Zeits. **18**, 533 (1917).

41. R. Fürth, Physik. Zeits. **20**, 303, 332, 350, 375 (1919); also reference 16.

42. H. Freundlich, Kapillarchemie (Leipzig, 1930–1932), Vols. **I** and **II**; see particularly pp. 485–510 in Vol. **I** and pp. 140–162 in Vol. **II**.

43. The Svedberg, *Die Existenz der Molekule* (Leipzig, 1912).

In reference 39 we have an extremely valuable account of the entire subject of Brownian motion and molecular fluctuations; there exists no better introduction to this subject than these lectures of Smoluchowski. In reference 40 Sommerfeld gives a fairly extensive bibliography of Smoluchowski's writings.

§*1*.—The theory of density fluctuations as developed by Smoluchowski represents one of the most outstanding achievements in molecular physics. Not only does it quantitatively account for and clarify a wide range of physical and physico-chemical phenomena, it also introduces such fundamental notions as the "probability after-effect" which are of very great significance in other connections (see Chapter IV).

44. M. v. Smoluchowski, Wien. Ber. **123**, 2381 (1914); see also Physik. Zeits. **16**, 321 (1915) and Kolloid Zeits. **18**, 48 (1916). For discussions of the problem of density fluctuations prior to the introduction of the notion of the "speed of fluctuations" see

45. M. v. Smoluchowski, *Boltzmann Festschrift* (1904), p. 626; *Bull. Acad. Cracovie*, p. 1057, 1907; Ann. d. Physik **25**, 205 (1908). Also

46. R. Lorenz and W. Eitel, Zeits. f. physik. Chemie **87**, 293, 434 (1914).

It is of some interest to recall that referring to his deviation of the formulae for $\langle\Delta_n\rangle_{Av}$ and $\langle\Delta_n^2\rangle_{Av}$ [Eqs. (356) and (358)] Smoluchowski says, "Aus diesem komplizierten Formeln [referring to the formula for $W(n;m)$] lassen sich mittels verwickelter summationen merkwurdigerweise recht einfache resultate fur die durschschnittliche Änderung der Teilchenzahl ableiten. . . . So wie fur das Anderungsquadrat bei unbestimmter Anfangszahl n [Eq. (363)]." This led to some heated discussion whether these formulae cannot be derived more simply; for example, see

47. L. S. Ornstein, Konink. Akad. Wetenschap. Amsterdam **21**, 92 (1917). But neither Ornstein nor Smoluchowski seems to have noticed that the formulae for $\langle\Delta_n\rangle_{Av}$ and $\langle\Delta_n^2\rangle_{Av}$ can be derived very directly from the fact that the transition probability $W(n;m)$ is the sum (in a technical sense) of a Bernoulli and a Poisson dis-

tribution; it is to this fact that the simplicity of the results are due.

§*2*.—Comparisons between the predictions of his theory with the data of colloid statistics were first made by Smoluchowski himself (reference 44). The experiments which were used for these first comparisons were those of

48. The Svedberg, Zeits. f. physik. Chemie **77**, 147 (1911); see also references 43 and 46. But precision experiments carried out with expressed intention of verifying Smoluchowski's theory are those of

49. A. Westgren, Arkiv for Matematik, Astronomi, och Fysik **11**, Nos. 8 and 14 (1916) and **13**, No. 14 (1918).

An interesting application of Smoluchowski's theory to a problem of rather different sort has been made by Fürth:

50. R. Fürth, Physik. Zeits. **19**, 421 (1918); **20**, 21 (1919). Fürth made systematic counts of the number of pedestrians in a block every five seconds. This interval of five seconds was chosen because the length of the block was such that a pedestrian observed in the block on one occasion has an appreciable probability of remaining in the same block when the next observation is made five seconds later. We can, accordingly, define a probability after-effect factor P $(=v\tau/a$, where v is the average speed of a pedestrian, τ the chosen interval of time and a the length of the block), and Smoluchowski's theory applies. A statistical analysis of this data showed that the agreement with the theory is excellent. It is amusing that by systematic counts of the kind made by Fürth it is possible actually to determine the average speed of a pedestrian!

§*3*.—The theory outlined in this section is derived from

51. M. v. Smoluchowski, Wien. Ber. **124**, 339 (1915); see also references 39 and 41.

§*4*.—Among the early discussions on the compatibility between the notions of conventional thermodynamics and the then new standpoint of the kinetic molecular theory, we may refer to

52. J. Loschmidt, Wien. Ber. **73**, 139 (1876); **75**, 67 (1877).

53. L. Boltzmann, Wien. Ber. **75**, 62 (1877); **76**, 373 (1877); also Nature **51**, 413 (1895) and *Vorlesungen über Gas Theorie* (Leipzig, 1895) Vol. **I**, p. 42 (or the reprinted edition of 1923).

54. E. Zermelo, Ann. d. Physik **57**, 485 (1896); **59**, 793 (1896).

55. L. Boltzmann, Ann. d. Physik **57**, 773 (1896); **60**, 392 (1897).

Smoluchowski's fundamental discussions of the limits of validity of the second law of thermodynamics are contained in

56. M. v. Smoluchowski, Physik. Zeits. **13**, 1069 (1912); **14**, 261 (1913). See also references 39 and 51.

It is somewhat disappointing that the more recent discussions of the laws of thermodynamics contain no relevant references to the investigations of Boltzmann and Smoluchowski [e.g., P. W. Bridgman, *The Nature of Thermodynamics* (Harvard University Press, 1941)]. The absence of references, particularly to Smoluchowski, is to be deplored since no one has contributed so much as Smoluchowski to a real clarification of the fundamental issues involved.

For an exhaustive discussion of the foundations of statistical mechanics, see

57. P. and T. Ehrenfest, *Begriffliche Grundlagen der Statistischen Auffassung in der Mechanik, Encyklopadie der Mathematischen Wissenschaften* (1911), Vol. 4, p. 4. And for Carathéodory's version of thermodynamics see

57a. S. Chandrasekhar, *An Introduction to the Study of Stellar Structure* (University of Chicago Press, 1939), Chap. I, pp. 11–37.

§*5*.—See Smoluchowski, reference 39; also

58. M. v. Smoluchowski, Ann. d. Physik **48**, 1103 (1915).

59. R. Fürth, Ann. d. Physik **53**, 177 (1917).

§*6*.—See Smoluchowski reference 39; also

60. M. v. Smoluchowski, Zeits. f. physik. Chemie **92**, 129 (1917).

61. R. Zsigmondy, Zeits. f. physik. Chemie **92**, 600 (1917). The papers 60 and 61 contain references to the earlier literature on the subject of coagulation. For the more recent literature see Freundlich (reference 42, particularly Vol. **II**, pp. 140–162).

§*7*.—See

62. H. A. Kramers, Physica **7**, 284 (1940). Also,

63. H. Pelzer and E. Wigner, Zeits. f. physik. Chemie, **B15**, 445 (1932).

An aspect of the theory of Brownian motion we have not touched upon concerns the natural limit set by it to all measuring processes. But an excellent review of this entire field exists:

64. R. B. Barnes and S. Silverman, Rev. Mod. Phys. **6**, 162 (1934).

Chapter IV

The ideas developed in this chapter are in the main taken from

65. S. Chandrasekhar, Astrophys. J. **94**, 511 (1941).

66. S. Chandrasekhar and J. von Neumann, Astrophys. J. **95**, 489 (1942).

67. S. Chandrasekhar and J. von Neumann, Astrophys. J. **97**, 1, (1943).

§*1*.—See references 65, 66, and 67; also

68. S. Chandrasekhar, *Principles of Stellar Dynamics* (University of Chicago Press, 1942), Chapters II and V.

§*2*.—The problem considered in this section is clearly equivalent to finding the probability of a given *electric* field strength at a point in a gas composed of simple ions. This latter problem was first considered by Holtsmark:

69. J. Holtsmark, Ann. d. Physik **58**, 577 (1919); also Physik. Zeits. **20**, 162 (1919) and **25**, 73 (1924). Among other papers on related subjects we may refer to

70. R. Gans, Ann. d. Physik **66**, 396 (1921).

71. P. Hertz, Math. Ann. **67**, 387 (1909).

72. R. Gans, Physik. Zeits. **23**, 109 (1922).

73. C. V. Raman, Phil. Mag. **47**, 671 (1924).

§*3*.—See references 66 and 67. See also three further papers on "Dynamical Friction" by Chandrasekhar in forthcoming issues of *The Astrophysical Journal* where further applications of the Fokker-Planck equation will be found.

REVIEWS OF MODERN PHYSICS VOLUME 15, NUMBER 1 JANUARY, 1943

The Diffraction of X-Rays by Liquid Elements

NEWELL S. GINGRICH
University of Missouri, Columbia, Missouri

I. INTRODUCTION

DISCOVERY of diffraction and interference effects associated with the passage of x-rays through matter very soon led to the recognition that studies of these effects are useful in determining the structure of matter. Work on the diffraction of x-rays by crystals was initiated in 1912 with the famous discovery made by Friedrich, Knipping, and Laue[1] and this field has been developed extensively[2] by many investigators to supply valuable information concerning the arrangements of atoms in crystals. Diffraction of x-rays by gases was first studied as early as 1911,[3] and subsequent work in this field[4] has supplied information not only on the arrangements of atoms in molecules, but also on the electron distributions within atoms. Early work on the diffraction of x-rays by liquids and noncrystalline substances dates back to 1913 when Friedrich[5] obtained x-ray diffraction patterns of Canada balsam, paraffin, and amber. In 1916, Debye and Scherrer[6] obtained diffraction patterns of several liquids, including benzene, and in 1922, Keesom and de Smedt[7] reported work of this type with liquid elements. Subsequent experimental investigations and theories in this field have made possible the determination of atomic distributions in liquids, giving results which are perhaps the most quantitative means at our disposal of describing the structure of a liquid.

Although x-ray diffraction patterns of many liquids have been studied, the present discussion is limited almost exclusively to liquid elements. The early work of Keesom and de Smedt[7] on nitrogen, oxygen, and argon supplied valuable qualitative information concerning the structure of liquid elements, but the theoretical advances made by Zernike and Prins[8] and by Debye and Menke[9] supplied a strong impetus to further study in this field because they offered a quantitative approach to the interpretation of experimental results. Within the decade following these theoretical advances, experimental studies were carried out on helium, lithium, nitrogen, oxygen, sodium, aluminum, phosphorus, sulphur, chlorine, argon, potassium, zinc, gallium, selenium, rubidium, cadmium, indium, tin, cesium, mercury, thallium, lead, and bismuth. In the cases of many of these elements, the available quantitative interpretation was applied, and in some cases, the effect of temperature on the liquid structure was investigated. For the two elements, argon and nitrogen, the effect of pressure was also investigated, including the case where the element was a permanent gas at high pressure.

Studies of x-ray diffraction by liquid elements have certain advantages over those with polyatomic liquids. In the first place, the theory by means of which the atomic distribution is obtained from the diffraction pattern, is more simply and accurately applied to an assemblage of atoms of one kind than to a polyatomic liquid, since fewer assumptions are required. In the second place, theoretical considerations concerning liquid structures and the nature of interatomic forces are more likely to deal with the simplest liquids than with the more complicated ones. This estimate has already been borne out, notably in the cases of argon, sodium, potassium, and mercury for which theoretical work has supplied interesting and valuable comparisons with, or deductions from, the experimentally determined atomic distribution functions. Finally, it is important to have as full and complete information as possible concerning the properties of all the elements.

II. THEORY

Although many investigators[10] have contributed to discussions of the theory of x-ray diffraction by liquids, the most fruitful approach to the present problem is that which had its origin with the early work of Debye[11] in which it was shown that one has always to consider two atoms whose scattered rays interfere with one another as a result of which an interference pattern is produced which is defined essentially by the relative separations of the two atoms. If the relative frequency with which various relative separations occurred were known completely, then the intensity patterns could be predicted. Zernike and Prins[8] proved in a one-dimensional model that certain distances receive special weight in the formation of interferences, and it is clear that this conclusion can be carried over, at least qualitatively, to three-dimensional liquids, since the impossibility of interpenetration of atoms and the existence of interatomic binding imply that certain arrangements of any given atom with respect to its neighbors are more probable than others. These investigators introduced the idea of a distribution function which, if known, would facilitate prediction of the x-ray pattern to be expected for any given substance. But even in the simplest case of atoms as hard spheres, it has not been possible to calculate unambiguously and exactly the distribution of atoms about any given atom, and hence this method can so far supply only qualitative results. If, however, the x-ray intensity curve is known precisely, then the reverse of the above procedure is available, and the distribution function can be determined from the experimental intensity curves. This was suggested by Zernike and Prins as a possibility, but it remained for Debye and Menke[9] to complete the analysis and to apply it for the first time to liquid mercury. An alternative presentation of this theory has been given by Warren and Gingrich[12] and by Warren,[13] and this is the one followed here.

Consider an incident x-ray beam of amplitude E_0 propagated in a direction defined by the unit vector S_0, polarized with its electric field normal to the plane of Fig. 1, and directed upon an atom at O. For a single electron at O, the amplitude of the radiation scattered to a point P at a large distance R from O and in a direction defined by the unit vector S is given by electromagnetic theory[14] as

$$E_p = E_0(e^2/mc^2R). \qquad (1)$$

If Z electrons were concentrated at O, then the amplitude of the scattered radiation at P would be Z times as great as for one electron. But for an atom of atomic number Z at O, the size of the atom is comparable to the wave-length of

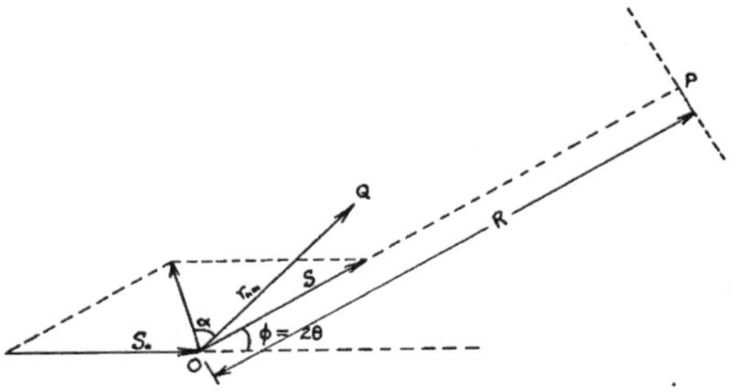

FIG. 1. Unit vector S_0 defines the direction of radiation incident upon an atom at O; unit vector S defines the direction of radiation scattered at angle ϕ; vector r_{nm} represents the interatomic displacement between atoms at O and Q.

x-rays, and hence there will be interference effects between the rays scattered by different electrons in the same atom. The resultant amplitude at P is then given by

$$E_p = E_0 f(e^2/mc^2R), \qquad (2)$$

where the factor f, called the atomic structure factor, is determined by the distribution of electrons in the atom, and is equal to, or less than Z. The phase of the radiation scattered from atom n at Q as compared to that from the origin atom is $2\pi/\lambda(r_n\cdot S_0 - r_n\cdot S)$, and thus, if there is an incident wave of the form $E=E_0 \exp(2\pi i\nu t)$, the scattered wave from atom n is given by

$$E_p = E_0(e^2/mc^2R)f_n \times \exp\left[2\pi i(\nu t - r_n\cdot(S_0-S)/\lambda)\right]. \qquad (3)$$

The total amplitude at P is the sum of all the amplitudes scattered to that point and is given by

$$E_p = E_0(e^2/mc^2R)\sum_n f_n \times \exp\left[\{2\pi i/\lambda\}(S-S_0)\cdot r_n\right] \qquad (4)$$

and the square of the amplitude of the radiation at P is obtained by multiplying (4) by its complex conjugate, giving

$$E_p^2 = E_0^2(e^4/m^2c^4R^2) \times\left[\sum_n f_n \exp(\{2\pi i/\lambda\}(S-S_0)\cdot r_n)\right] \times\left[\sum_m f_m \exp(\{2\pi i/\lambda\}(S-S_0)\cdot r_m)\right]. \qquad (5)$$

Removing the restriction of plane polarized radiation introduces the polarization factor derived elsewhere[14] and we have

$$E_p^2=\{(1+\cos^2\phi)/2\} \times E_0^2(e^4/m^2c^4R^2)\sum_n\sum_m f_nf_m \times\exp[\{2\pi i/\lambda\}(S-S_0)\cdot(r_n-r_m)]. \quad (6)$$

The coefficient of the double summation is proportional to the intensity scattered by a single electron, represented by I_e, and E_p^2 is proportional to the intensity of scattered radiation I. Furthermore, $r_n-r_m=r_{nm}$, $S-S_0=2\sin\theta$, where θ is half the scattering angle, s is defined as $(4\pi\sin\theta)/\lambda$, and α is the angle between $(S-S_0)$ and r_{nm}. Then if I/I_e is designated I_{eu}, intensity in electron units, Eq. (6) may be written as

$$I_{eu}=\sum_n\sum_m f_nf_m\exp(isr_{nm}\cos\alpha). \quad (7)$$

Now, as the vector r_{nm} takes on all positions in space, α takes on all values at random, and the effect of this random orientation may be obtained by averaging the exponential function over all solid angle. Thus

$$\int_0^{4\pi}[\exp(isr_{nm}\cos\alpha)]d\Omega\Big/\int_0^{4\pi}d\Omega =\int_0^{\pi}[\exp(isr_{rm}\cos\alpha)]2\pi\sin\alpha d\alpha/4\pi =(\sin sr_{nm})/sr_{nm}$$

and hence we have Debye's equation

$$I_{eu}=\sum_n\sum_m f_nf_m\{(\sin sr_{mn})/sr_{mn}\}. \quad (8)$$

It is seen that the intensity depends upon the structure factors of the atoms, the angle of scattering, the wave-length of the x-rays, and upon the interatomic distances r_{nm}. In the present discussion for atoms of one kind, $f_n=f_m=f$, and $f_nf_m=f^2$.

In performing the summations, it is necessary to consider one atom and to carry out the summation over all distances to all atoms of the array, including the origin atom, and then to allow the origin atom to be every atom in the array in turn. Summations for any atom with respect to itself lead simply to unity in each case since in the limit as $r_{nm}\rightarrow 0$, $\sin sr_{nm}/sr_{nm}\rightarrow 1$. Thus, if there are N atoms in the array,

$$I_{eu}=Nf^2[1+\sum_n{}'(\sin sr_{nm})/sr_{nm}], \quad (9)$$

where this summation excludes the origin atom. If it is assumed that there is a continuous distribution of atoms, then the above summation may be replaced by an integral. If $\rho(r)$ is the density of atoms at distance r from the origin atom, then the number of atoms in a spherical shell of radius r and thickness dr is $4\pi r^2\rho(r)dr$, and this is sometimes referred to as the radial density of atoms, or the atomic distribution function.

Equation (9) is then written as

$$I_{eu}=Nf^2\left[1+\int_0^R 4\pi r^2\rho(r)\{(\sin sr)/sr\}dr\right], \quad (10)$$

where R is the radius of the liquid sample, which is very large compared to any value of r of interest here. If we take ρ_0 as the constant average density of atoms, then Eq. (10) may be rewritten

$$I_{eu}=Nf^2\left[1+\int_0^R 4\pi r^2(\rho(r)-\rho_0)\{(\sin sr)/sr\}dr +\int_0^R 4\pi r^2\rho_0\{(\sin sr)/sr\}dr\right]. \quad (11)$$

The second integral may be taken as zero, under normal conditions, for physical reasons. It is seen that

$$\int_0^R 4\pi r^2\rho_0\{(\sin sr)/sr\}dr=\Big[(4\pi r^3/3)\rho_0(3/(sr)^2) \times(\{(\sin sr)/sr\}-\cos sr)\Big]_{r=0}^{r=R}. \quad (12)$$

For $r=0$, this is zero regardless of s, and for $r=R$, a very large quantity, the integral is zero unless s is extremely small. For s corresponding to a scattering angle of the order of minutes or seconds of arc, this integral should contribute to the intensity pattern, but this portion of the intensity pattern is wholly unobservable because of the presence of the main beam. For normal sample sizes, and as long as we limit ourselves to observable intensity, this integral may then be taken as zero.

Equation (11) may be rearranged into a more useful form by writing

$$i(s)=(I_{eu}/Nf^2)-1 \quad (13)$$

and by taking the limits as 0 to ∞ instead of 0 to R. This change of limit is necessary for later application of the Fourier integral theorem, and it is justified by the fact that $\rho(r)$ rapidly approaches ρ_0 as r reaches several atomic diameters, which is still many orders of magnitude less than R. Then

$$si(s)=\int_0^\infty 4\pi r(\rho(r)-\rho_0)\sin srdr, \qquad (14)$$

which is transformed by the Fourier integral theorem into

$$r\{\rho(r)-\rho_0\}=(1/2\pi^2)\int_0^\infty si(s)\sin rsds \qquad (15)$$

or

$$4\pi r^2\rho(r)=4\pi r^2\rho_0+(2r/\pi)\int_0^\infty si(s)\sin rsds. \qquad (16)$$

Equation (16) is the one which has been used extensively in the determination of the distribution function $4\pi r^2\rho(r)$ giving the number of atoms between r and $r+dr$ from any arbitrary atom within the liquid element. Two important steps remain to complete this determination; first, the evaluation of the function $i(s)$ from experiment, and second, the evaluation of the integral of Eq. (16).

III. EXPERIMENTAL

Inspection of the theoretical development shows that the following experimental conditions are called for: (1) monochromatic x-rays, (2) well-defined directions of incidence and scattering, (3) a true measure of the coherently scattered intensity from the liquid element as a function of s over a very wide range of s, (4) no distortion in the relative intensity measurements due, for example, to polarization or absorption effects, (5) the same units for I and Nf^2 of Eq. (13). These are some of the more important conditions to be considered and met as closely as possible.

A. Radiation

Radiation from an x-ray tube target is made up of continuous radiation and characteristic (or line) radiation and two methods have been used to attempt to satisfy condition (1) above, first, crystal reflection, and second, selective filtering. Generally, high intensities have been attained with the second method by using a band of wave-lengths, but this does not satisfy condition (1) Crystal reflection does supply monochromatic radiation as long as the voltage on the x-ray tube is lower than that which is necessary to excite continuous radiation of one-half the wave-length of the desired characteristic radiation. Furthermore, it has been shown[15] that improper filtering can give rise to spurious diffraction peaks, and that filtering to obtain monochromatic radiation closely approximate to that obtained by crystal reflection is not as efficient as crystal reflection itself in the matter of exposure time. Although several different crystal monochromators have been tried, rocksalt appears to be the most suitable for this purpose.

The choice of characteristic radiation to be used is determined largely by considerations of available targets, absorption in the sample, and the region of s to be emphasized. Thus, for detailed investigation of scattering at small values of s, long wave-lengths such as that of Cu are useful. For emphasis on scattering at large s, a short wave-length such as that of Ag is useful, and for general purposes, Mo radiation is quite suitable.

Unpolarized radiation from the x-ray target undergoes partial polarization when scattered by the liquid sample, and also when reflected from the crystal monochromator if this is used. It is shown elsewhere[14] that the polarization factor is $(1+\{\cos^2 2\theta\}/2)$ for scattering from the sample. If 2ψ is the total angle of scattering from the crystal, it can be proved that this factor should be $\{1+(\cos^2 2\psi \cos^2 2\theta)/2\}$. That this is so can be seen qualitatively from the fact that only the component of the electric field (in the electromagnetic x-radiation) normal to the plane of scattering remains unaltered, while the component in the plane of scattering may suffer a diminution at each scattering. Since scattered intensity with no effect of polarization is desired, the experimental intensity is to be divided by whichever of the above two factors is appropriate under particular experimental conditions.

B. Detectors of Radiation

Photographic film has been the most commonly used agency for detection of radiation in this field of work. Cylindrical cameras are used, with the sample at the center of the cylinder, and with the film on the circumference. For accurate measurement of diffraction angles, a fairly large camera, of perhaps 9-cm radius, is desirable. It has been demonstrated[16] that (1) ordinary x-ray film responds faithfully for reasonable densities of exposure, that (2) double-emulsion type film is considerably more effective than ordinary films for Mo radiation, and that (3) the use of intensifying screens for the measurement of relative intensities introduces almost insurmountable complications. Careful microphotometering of the films is, of course, a necessary adjunct to this method.

Electrical methods of detection are available, and in this category, a Geiger-Mueller counter[17] has been used successfully for x-ray diffraction by a liquid element. Considerable preparation must be made for this method, but where temperatures and pressures of the sample must be accurately maintained, the shorter time required to obtain a complete pattern, as compared with the photographic method, may justify this preparation.

C. The Sample

Many different types of samples have been used to meet the special requirements of the various liquid elements. In the case of He, no container was used, but a free stream of the liquid constituted the sample. Unfortunately it is not practical to adopt this ideal scheme for many elements. Cylindrical glass tubes of very thin walls have been used, with diameters ranging from about 0.05 cm to 0.3 cm and with wall thicknesses ranging between 0.002 cm and 0.01 cm in various cases. For Li, a cell of triangular section in a horizontal plane was used with a maximum thickness of 1.8 cm. In this case, the size of the sample constituted an appreciable fraction of the camera radius, and hence the pattern was probably somewhat smeared out because of the poor geometrical conditions. In some instances, flat cells were used, with the incident beam making an angle of 90° or less with the plane of the cell. With this arrangement, cell thicknesses were from 0.17 cm to 0.008 cm, with mica windows 0.0005 cm thick, or with Be windows 0.03 cm to 0.07 cm thick. In addition to the above schemes for transmission, diffraction patterns have also been obtained by reflection from liquid surfaces, as was done in the case of mercury.

1. Optimum Thickness

The transmission method has been used most frequently, and in this case there is an optimum thickness of cell, since a cell which is too thin

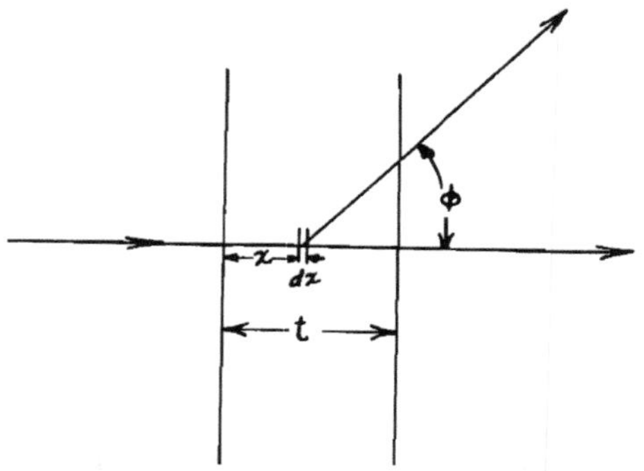

FIG. 2. Representing scattering at an angle ϕ from a layer of thickness dx at a depth x in a flat sample of thickness t. The incident beam is perpendicular to the plane of the sample.

has insufficient scattering matter, and since one of great thickness absorbs too strongly. Consider the case of a flat sample of thickness t and linear absorption coefficient μ with an incident beam of intensity I_0 directed perpendicularly upon the sample, as in Fig. 2. Then a thickness dx at depth x will contribute to the intensity scattered at angle $\phi = 2\theta$ in the amount dI_ϕ where

$$dI_\phi = [cI_0 \exp(-\mu x)dx](\exp[-\mu(t-x)/\cos\phi])$$

or

$$I_\phi = K[\exp(-\mu t) - \exp(-\mu t/\cos\phi)],$$

and

$$K = (cI_0 \cos\phi)/\mu(1-\cos\phi).$$

Applying the condition for optimum thickness,

$$dI_\phi/dt = K[-\mu \exp(-\mu t_0) + (\mu/\cos\phi)\exp(-\mu t_0/\cos\phi)] = 0$$

and hence

$$t_0 = (1/\mu)[-(\cos\phi \log\cos\phi)/(1-\cos\phi)], \quad (17)$$

which, in the limit as ϕ approaches 0, is $t_0 = 1/\mu$. Thus, for scattering at small angles, the optimum thickness is the reciprocal of the linear absorption coefficient, but for other scattering angles, it is less than $1/\mu$. Figure 3 shows the optimum thickness as a function of scattering angle ϕ. If the plane of the flat sample is not perpendicular to the incident beam, then the above relation should be modified to meet this condition. For a cylindrical sample completely bathed in incident radiation, the expression $t_0 = 1/\mu$ should be a good approximation to the optimum diameter for very small values of ϕ, but experience shows that a cylinder of diameter as much as two or three times t_0 gives more intense patterns than one of t_0. This is presumably due to the fact that one is generally interested in the diffraction pattern at points considerably removed from the incident beam, for which the above expression for optimum thickness is not valid.

2. *Absorption Corrections*

Inasmuch as the rays scattered at different angles suffer different amounts of absorption in the sample itself, the true relative intensity of one portion of the pattern with respect to that of another can be obtained only after correction for this differential absorption.

Consider first, the case of a flat sample of thickness t with the normal to its surface making an angle σ with the direction of the incident beam as shown in Fig. 4. If the incident beam is

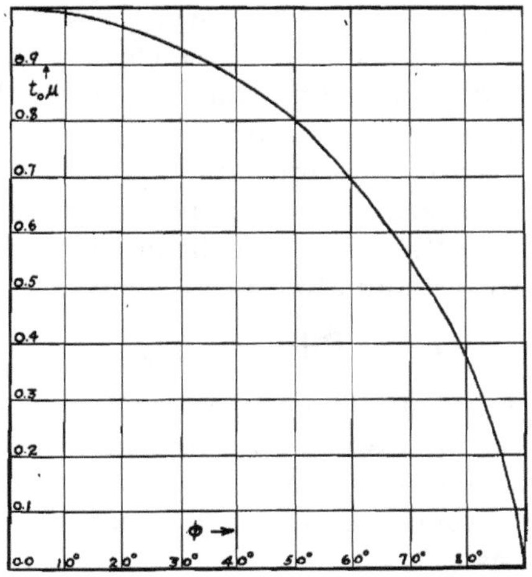

FIG. 3. Showing how optimum thickness of a flat sample varies with the scattering angle.

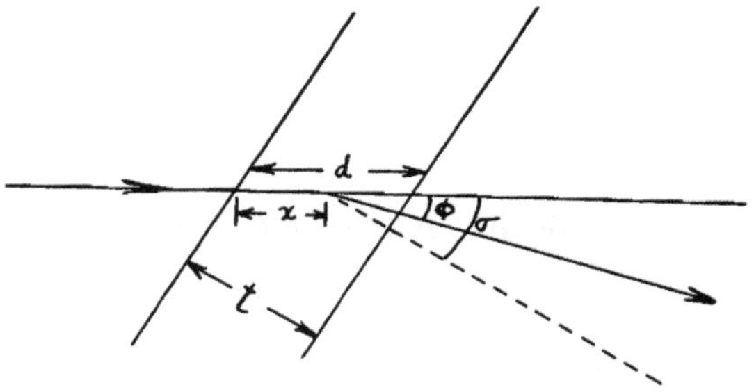

FIG. 4. Directions and angles defined for the computation of the absorption correction in a flat sample whose normal is inclined at an angle σ with the direction of incident radiation.

of original intensity I_0, the contribution to the intensity scattered at angle ϕ by a thickness dx at depth x in the sample is

$$dI_\phi = K[\exp(-\mu x)dx] \times \{\exp[-\{\mu(d-x)\cos\sigma\}/\cos(\sigma-\phi)]\}, \quad (18)$$

where K includes I_0 and other constant factors. Then

$$I_\phi = K \exp[-\mu d \cos\sigma/\cos(\sigma-\phi)] \int_0^d \exp\{\mu x[\cos\sigma - \cos(\sigma-\phi)]/\cos(\sigma-\phi)\}dx, \quad (19)$$

or

$$I_\phi = K/\mu z \cos\sigma \times \{\exp[-\mu d\cos\sigma/\cos(\sigma-\phi)]\} \times [\exp(\mu z d\cos\sigma) - 1], \quad (20)$$

where

$$z = [\cos\sigma - \cos(\sigma-\phi)]/\cos\sigma\cos(\sigma-\phi).$$

Since relative intensity is desired, it is necessary to normalize the above expression with respect to some particular direction. We refer the intensity at any angle ϕ to that at $\phi = 0$. From Eq. (19) it is seen that for $\phi = 0$,

$$I_{\phi=0} = Kde^{-\mu d} \quad (21)$$

and, noting that $t = d\cos\sigma$, we have

$$I_\phi/I_{\phi=0} = (1/\mu tz)[1 - e^{-\mu tz}]. \quad (22)$$

This relation may be used with the sample perpendicular to the main beam, for in this case, $\sigma = 0$ and $z_\perp = (1-\cos\phi)/\cos\phi$. When $\sigma \neq 0$, an alternative procedure is to normalize with respect to the intensity at $\phi=\sigma$. Setting $\phi=\sigma$ in Eq. (20) and taking the ratio of I_ϕ to $I_{\phi=\sigma}$, we have,

$$I_\phi/I_{\phi=\sigma}=\{(\cos\sigma-1)/\cos\sigma[\exp(-\mu t)-\exp(\mu t\cos\sigma)]\}\{[\exp\{-\mu t/\cos(\sigma-\phi)\}]\times[\exp(\mu z t)-1]/z\}. \quad (23)$$

The experimental intensity is to be corrected by dividing it by Eq. (22) or Eq. (23), and this correction can be made along with the polarization correction described above. This corrected intensity is that which one would observe if there were no effect of absorption or polarization.

In the reflection method, assume that all rays strike the liquid surface at an angle α. A volume element $(Adx)/\sin\alpha$ at a depth x below the surface will be irradiated by primary rays which have traveled a distance $x/\sin\alpha$ in the liquid, and a secondary ray scattered at an angle 2θ with the primary beam travels through the liquid a distance $x/\sin(2\theta-\alpha)$ before reaching the surface. The intensity of rays scattered at angle 2θ is therefore

$$I_{2\theta}=K(1/\sin\alpha)\times\int_0^\infty \exp\{-\mu[x/\sin\alpha+x/\sin(2\theta-\alpha)]\}dx, \quad (24)$$

or

$$I_{2\theta}=(K/\mu)[\{\sin(2\theta-\alpha)\}/\{\sin\alpha+\sin(2\theta-\alpha)\}]. \quad (25)$$

Relating the intensity at any angle to that at $2\theta=2\alpha$

$$I_{2\theta}/I_{2\theta=2\alpha}=(1/2)\{[\sin(2\theta-\alpha)]/[\sin\alpha+\sin(2\theta-\alpha)]\}. \quad (26)$$

For a cylindrical sample completely bathed in a primary beam, the absorption correction has been determined by means of a graphical integration and the results are plotted elsewhere for convenient use.[18] This correction is generally smaller for large angle diffraction than that for the flat sample, and hence the cylindrical sample is preferable to the flat sample in this respect. On the other hand, this correction has been computed on the assumption that the cylindrical sample is completely bathed in incident radiation of uniform intensity across the entire section of the beam, and this is seldom true for crystal monochromated x-rays. It has been shown,[19] however, that no marked change of relative intensity in the diffraction pattern of liquid sodium occurred when several diameters of sample were used, from about two to five times the width of the incident beam.

Use of a crystalline material such as mica for the sample holder has an advantage in that the diffraction produced by it is largely concentrated in sharp spots and hence can easily be distinguished from the true liquid pattern, whereas a glass holder gives a pattern closely similar to that of liquids and hence cannot easily be distinguished from the desired pattern. But the choice of construction for a sample holder is not a simple matter, since considerations of strength, chemical properties, and many other features are also important.

3. *Incoherent Radiation and Curve Fitting*

In determining the function $i(s)$ from experimental data, it is obviously necessary to have I_{eu} and Nf^2 in the same units. This is done by assuming that at large angles where interference effects can no longer be observed in the diffraction pattern, the observed coherently scattered x-rays have the same intensity as that produced by the same number of atoms which scatter x-rays independently of one another with no interference effects. If the atoms scatter x-rays independently, then the curve for the coherent scattering of N atoms will be of the same form as that for a single atom. These curves can be obtained from tables[20] of atomic structure factors. But before the above fitting of curves can be made, a correction must be applied to the observed intensity curve for incoherent radiation. It is shown elsewhere that the intensity of x-rays scattered from an atom is

$$I=I_e[f^2+R(Z-\sum f_n^2)], \quad (27)$$

where I_e is the Thompson scattering per electron, f is the atomic structure factor, f_n is the structure factor of the nth electron in the atom, Z is the atomic number, and R is a recoil factor defined as

$$R=1/(1+\{h/mc\lambda\}\text{ vers }\phi)^3.$$

In the above expression for I, the term $I_e f^2$ is identified with coherent scattering and the term $I_e R(Z-\Sigma f_n^2)$ is identified with incoherent scattering which plays no part in interferences, but which introduces a slowly varying background on all patterns. The incoherent scattering function is conveniently tabulated elsewhere.[21] The following intensity ratios are useful in the curve-fitting process:

Incoherent/Total

$$=[R(Z-\sum F_n^2)]/[f^2+R(Z-\sum f_n^2)], \quad (28)$$

$$\text{Incoherent/Coherent}=[R(Z-\sum f_n^2)]/f^2, \quad (29)$$

$$\text{Coherent/Total}=f^2/[f^2+R(Z-\sum f_n^2)]. \quad (30)$$

Thus, for example, if the observed intensity has been corrected for polarization and absorption, and if at some large angle of scattering, there is no evidence of interference peaks, then the amount of incoherent radiation at that point can be obtained from Eq. (28). The difference between the total and the incoherent will be the coherent. Knowing the coherent intensity at this point enables one, from the f^2 tables, to draw in the independent coherent scattering at all angles. From this, and Eq. (29), the incoherent scattering at all angles can be drawn in, and subtracted from the observed intensity at all points, thus giving the I_{eu} to be used in the function $i(s)$, along with the coherent or Nf^2 curve. It is perhaps clearer to note that what has been done is to use

$$[(I_{eu}/N)/f^2]-1 \text{ for } i(s).$$

The incoherent correction for elements of high atomic number is very small but for elements of low atomic number, it is quite large. When the incoherent correction is small, the recoil factor may be taken as unity without introducing appreciable error.

One important requirement of the theory is that the experimental curve be known accurately to very large values of s, where $s=\{4\pi \sin\theta\}/\lambda$. This is important for two reasons, first, because all the interference effects should be included in the analysis, and second, because it is essential that the fitting shall be made at a point where the scattering is as completely independent as possible. Experimentally, it has not been possible, because of the weakness of the radiation, to carry the experimental observations much beyond $s=12$, but it is possible that future refinements will be in the nature of improved experimental technique to extend the range of s beyond that which is now practicable. It has been shown that in one particular case,[22] slight extension of this range has not markedly altered the results obtained, although in other cases, extension of the range has had considerable effect on the final result.

Because of the low intensity of coherent radiation at large angles where this curve-fitting is performed, much uncertainty in the whole analysis can be introduced at this step unless special attention is given to careful intensity determination at large angles. One partially successful attempt[23] has been made to minimize reliance upon the exact determination of weak coherent intensity at one value of the angle of diffraction, as is usually done. In effect, it is an attempt to perform the curve-fitting process over the whole curve rather than at one point.

4. *The Analysis*

After the intensity function $i(s)$ has been determined, the remaining problem is to evaluate the integral in the expression

$$4\pi r^2\rho(r)=4\pi r^2\rho_0+(2r/\pi)\int_0^\infty si(s)\sin rs\,ds. \quad (31)$$

Three methods have been used for this purpose: (1) the use of an harmonic analyzer, (2) graphical integration, and (3) trigonometric interpolation. A brief description of these three methods will be given.

Consider, first, for example, the use of the Coradi analyzer. If r is taken as a multiple of some chosen constant r_0, then $r=nr_0$ where n is the order of a harmonic and, if the function $si(s)$ is expressed as a sine series,

$$si(s)=\sum_0^\infty A_n \sin nsr_0;$$

then it can be shown[24] that

$$A_n=(1/\pi)\int_0^{2\pi} si(s)\sin nr_0 s\,ds$$

and the Coradi analyzer gives nA_n directly as the reading of the instrument after tracing over the $si(s)$ curve with gears in place for harmonic n. The ordinate scale for the $si(s)$ curve may be chosen for any convenient use with the instrument and the appropriate linear scale factor must then be applied to the reading obtained. Choice of the abscissa scale is determined by the construction of the analyzer and by the constant r_0. It is desired that the range of the argument, sr_0n, shall be equal to the total range of abscissa of the instrument, $2\pi n$, which covers 40 cm in the usual type of Coradi analyzer. Thus the limiting value of s, designated s_0, is related to r_0 by the relation $s_0r_0n=2\pi n$ or $s_0=2\pi/r_0$. If points on the distribution curve are desired at intervals of 0.4A, $r_0=0.4$, or $s_0=15.7$ and hence we have 40/15.7 or 2.55 cm per s unit. Plotting the abscissa as 2.55 s will then supply 0.4A per harmonic. Similarly, by use of 2.80 s we obtain 0.44 per harmonic. Gears could be constructed to supply fractional harmonics, or the existing gears can be interchanged from one range to another to give fractional harmonics. Thus, for example, over certain intervals, this procedure has supplied data for every 0.1A. Integration limits from 0 to 2π are required by the instrument rather than the desired limits of 0 to ∞, but since the function $si(s)$ is zero beyond the point of curve-fitting, as far as experiment is able to show, and since r_0 is chosen to include all the available data in the range of abscissa of the instrument, then this limited range of integration is permissible.

The integral of Eq. (31) can be evaluated by means of graphical integration. If $\beta=nr_0s$ and r_0 is taken as 0.4A, then β (in degrees) $=22.92ns$, and cards can be made listing sin β as a function of s for any n which is desired. Multiplying $si(s)$ and sin β, plotting this product as a function of s and planimetering the area under this curve gives the required result for one value of r. In this method, the number of curves to be computed, plotted, and planimetered is the same as the number of points in the distribution curve. A variation of this scheme is to plot $si(s)$ against s (or sin θ/λ for further simplification), draw the reflection of this curve to obtain an envelope, prepare cards for each harmonic with the abscissae indicated where the sine function takes on values of 0, 0.5, 0.866, 1.0 over a complete cycle, determine $si(s)$ sin nr_0s from these cards and the envelope, and finally planimeter as before. In this way, several determinations can be made within one envelope.

The method of trigonometric interpolation has been used by Danielson and Lanczos[25] to evaluate the integral of Eq. (31). This method depends on expressing the $si(s)$ curve in an approximate analytical form, a finite trigonometric series, chosen in such a way that the integral for each of a chosen set of values of r becomes simply related to one of the coefficients of the series. The s axis of the $si(s)$ curve is divided into some number, say 36, equal intervals. Arithmetic manipulation of the ordinates at these points of the $si(s)$ curve, together with a companion list of sine functions, gives the desired result. Prepared forms[25] greatly facilitate this procedure.

IV. RESULTS

1. Helium

Working under exceedingly difficult experimental conditions, Keesom and Taconis[26] passed Cu $K\alpha$ filtered radiation through free streams of liquid He I (-270.5°C) and He II (-271.4°C). Stream diameters of 1.5 mm and 3.0 mm were used, and in all cases the diffraction patterns had dense backgrounds which presumably were due largely to incoherent radiation. In both He I and He II, the position of the one observed peak was determined to be sin $\theta/\lambda=0.157$. No complete analysis was attempted, to obtain a distribution curve, but an analogy was made between the observed pattern and that to be expected from smeared-out crystal models, following the method of Prins and Petersen.[27] It was concluded that the liquid structure was more nearly similar to that of a face-centered cubic type than to that of a diamond-like structure.

2. Lithium

This element (m.p. 186°C) was investigated at 200°C by Gamertsfelder.[28] In a cylindrical camera of 9.20-cm radius, the sample had a horizontal cross section in the shape of a right isosceles triangle with the incident, crystal-reflected, Mo $K\alpha$ radiation passing 1.8 cm through the liquid along the altitude of the triangle. Because of the large

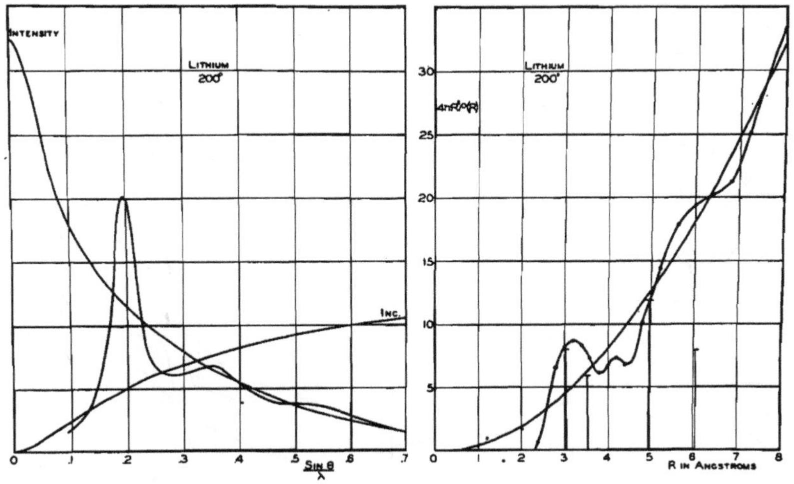

FIG. 5. Intensity curve and distribution curve for liquid lithium at 200°C. Ideal crystalline distribution out to 6A shown by vertical lines.

sample size relative to that of the camera, considerable blurring of the pattern is to be expected, and hence some uncertainty is introduced in the distribution curve on this account. Very thin mica windows were used to minimize undesirable scattering. Figure 5 shows the intensity curve plotted to an arbitrary ordinate scale, and the distribution curve, together with vertical lines to represent the ideal distribution in the crystalline form. The unusually large incoherent correction to the observed intensity is noted on the intensity graph of Fig. 5. The first peak in the distribution curve occurs at about 3.24A and it represents a concentration of roughly 9.8 atoms. The small second peak at 4.2A is not correlated closely with any distance in the crystal.

3. Nitrogen

Liquid nitrogen has been studied by Keesom and de Smedt[7] and by Sharrah[29] and nitrogen gas under high pressures has been studied by Harvey.[30] In the early work of Keesom and de Smedt, peaks in the intensity pattern were reported at $\sin\theta/\lambda = 0.139$ and 0.208, but in the recent work of Sharrah, who used monochromatic Mo $K\alpha$ radiation, intensity peaks were found at $\sin\theta/\lambda = 0.144$, 0.26, and 0.42 for nitrogen at 89°K, and at 0.148, 0.26, and 0.42 for 64°K. Figure 6 shows the results obtained by Sharrah. The first intensity peak is very prominent, whereas the other peaks are quite weak. Peaks in the distribution curve are at nearly the same distances for both temperatures: 1.3A, 2.6A, 4.0A, and 4.8A. The areas under the first peaks are 1.03 for 89°K and 1.08 for 64°K, and these peaks are discrete, implying permanent neighbors. Within the precision of this determination, the number of nearest neighbors is one, which confirms the existence of the N_2 molecule in liquid nitrogen. Harvey[30] showed that for compressed nitrogen gas, the x-ray scattering at small angles decreases as the pressure is increased.

4. Oxygen

Keesom and de Smedt[7] observed intensity peaks in the pattern from liquid oxygen, at $\sin\theta/\lambda = 0.154$, 0.24, and 0.35. Sharrah and Gingrich[31] more recently have used monochromatic radiation and they report intensity peaks at $\sin\theta/\lambda = 0.157$, 0.35, and 0.5 for the 89°K case, and at 0.159, 0.35, and 0.5 for the 62°K case. Their intensity curves are shown in Fig. 7, together with the distribution curves for the two temperatures. Peaks in these latter curves occur at 1.3A, 2.2A, 3.4A, and 4.2A for 89°K, and at 1.25A, 2.15A, 3.2A, and 4.1A for 62°K. The area under the first 89°K peak is 1.08 while that for 62°K is 1.18. Considering experimental precision, it is believed that the 1.08 is not significantly different from unity, and it is somewhat doubtful

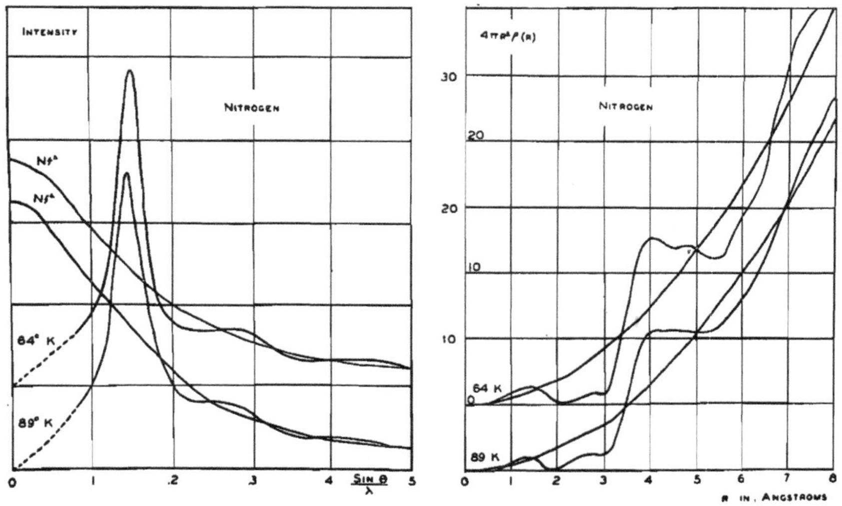

FIG. 6. Intensity curves and distribution curves for liquid nitrogen at 64°K and 89°K.

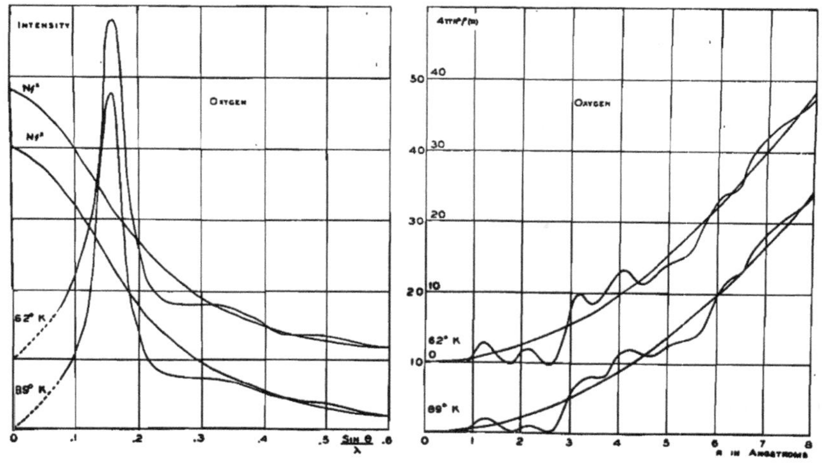

FIG. 7. Intensity curves and distribution curves for liquid oxygen at 62°K and 89°K.

whether the 1.18 should be interpreted as giving unambiguous proof of the existence of slightly more than one atom, on the average, at this distance. At both temperatures, a second discrete peak persisted in all determinations, a feature which is unique in liquid elements so far studied, to liquid oxygen. The occurrence of this second, discrete peak, and the excess areas under the first peak are suggestive of the presence of O_3 in liquid oxygen, since in this form, some atoms would have two nearest neighbors at about 1.3A, and some other atoms would have one nearest neighbor at 1.3A and one neighbor at 2.2A which is the approximate distance of the observed second peak. Existence of the third and fourth peaks have not been cited as clear-cut evidence for the presence of larger aggregates of oxygen atoms.

5. Sodium

Work on liquid sodium (m.p. 97.5°C) has been reported by Keesom,[32] Randall and Rooksby,[33] Tarasov and Warren,[34] and Trimble and Gingrich.[35] Quantitative work using monochromatic radiation was attempted in the last two cases, and their results are in essential agreement. Figure 8 shows the intensity curves and the distribution curves. Intensity peaks occur at $\sin\theta/\lambda=0.162$, 0.29, and 0.42 for 100°C and at 0.157, 0.28, and 0.4 for 400°C. The first peak in the distribution curve is at about 3.83A for 100°C and at about 3.90A for 400°C. These two distribution curves give an excellent illustration of the smearing-out effect on the distribution curve due to increased thermal agitation of the atoms for the liquid at high temperatures.

Using a simplified model of the liquid state, Wall[36] has calculated with considerable success the latent heats of fusion and of vaporization for sodium from the distribution curves given here.

6. Aluminum

Randall and Rooksby[37] and Gamertsfelder[28] have reported work on liquid aluminum (m.p. 659°C). In the more recent work,[28] the sample was simply an aluminum wire heated by means of a gas flame, with the very thin oxide coat acting as the sample holder. Figure 9 shows the intensity and the distribution curves for 700°C.

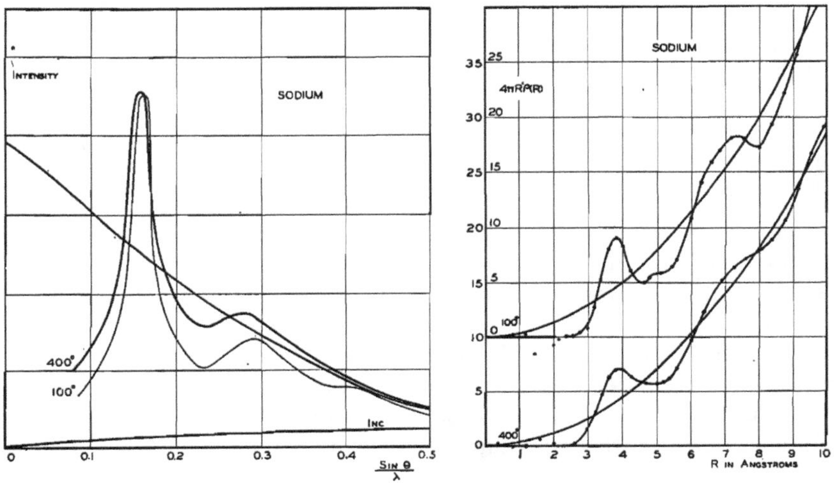

FIG. 8. Intensity curves and distribution curves for liquid sodium at 100°C and 400°C.

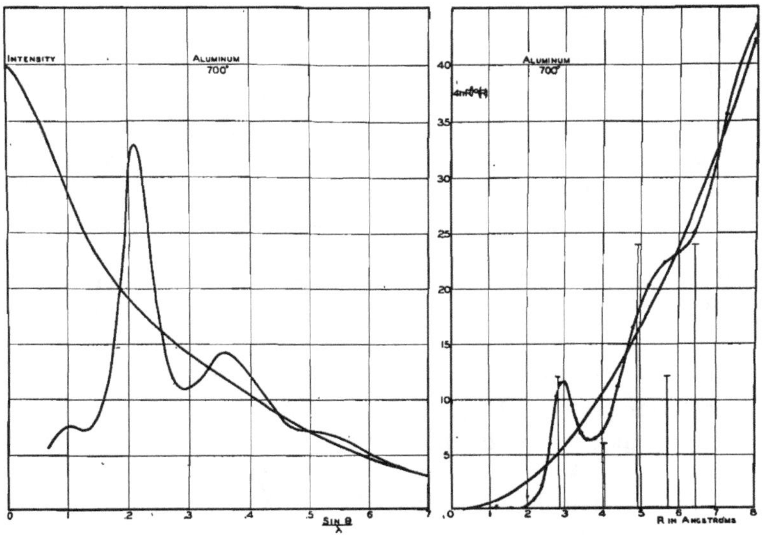

FIG. 9. Intensity curve and distribution curve for liquid aluminum at 700°C.

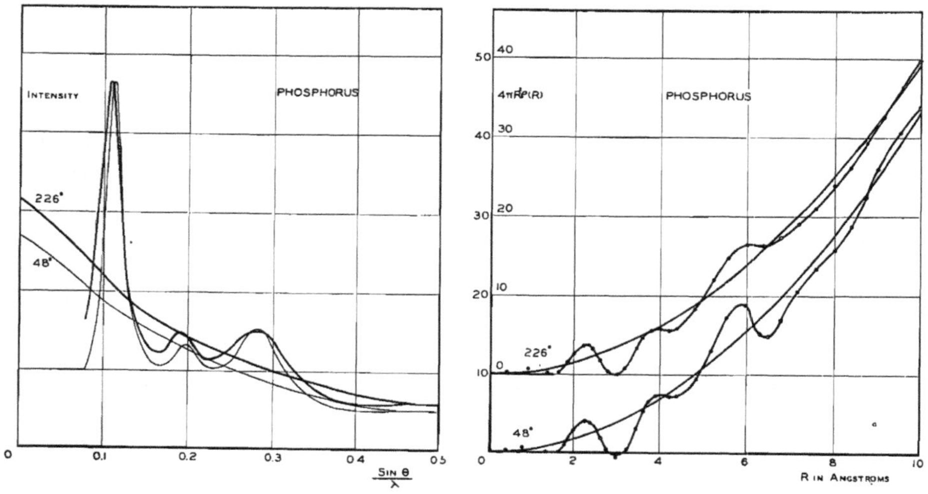

FIG. 10. Intensity curves and distribution curves for liquid yellow phosphorus at 48°C and 226°C.

In the intensity curve, a very weak peak appeared at $(\sin\theta)/\lambda = 0.10$, and the other peaks, whose intensities fall off in roughly the usual manner, occur at 0.210, 0.36, and 0.55. The first peak in the distribution curve is at 2.96A, it is not discrete, and the area under it represents about 10.6 atoms. It is seen that in this case, there is a fair correlation with the structure of the crystal. As in other cases where the first distribution peak is not discrete, the nearest neighbors are constantly changing, and what one gets is an average number of nearest neighbors, with frequent interchanging of these neighbors.

7. Phosphorus

The only work on liquid yellow phosphorus (m.p. 44.1°C) was reported by Thomas and Gingrich.[38] Their intensity and distribution curves for temperatures 48°C and 226°C are shown in Fig. 10. Intensity peaks occur at $(\sin\theta)/\lambda = 0.111$, 0.198, 0.28, for 48°C, and at 0.108, 0.194, and 0.28 for 226°C. Peaks in the distribution curves are found at very nearly the same distances for both temperatures, 2.25A, 3.9A, and 6.0A. Of particular interest here, is the fact that the first of these peaks at both temperatures is discrete, and represents approximately three atoms, thus showing that in liquid yellow phosphorus, each atom has three permanent nearest neighbors. This confirms, very directly, the existence of P_4 molecules in liquid yellow phosphorus. Rushbrooke and Coulson[39] have identified the second and third peaks with frequently recurring distances between atoms which are not in the same P_4 molecule.

Both red and black phosphorus occur in an amorphous form, and the atomic distribution curve has been determined in these cases.[23, 38] Here again there are three permanent nearest neighbors at a distance (2.28A) slightly (perhaps insignificantly) greater than that in liquid yellow phosphorus (2.25A). Beyond the first peak, however, the distribution curves differ considerably.

8. Sulphur

Blatchford,[40] Das,[41] Das and Das Gupta,[42] Prins,[43] and Gingrich[44] have reported work on this element (m.p. 113°C or 119°C). Although interesting information has been supplied in all cases, a complete analysis has been given in only one case,[44] for which the intensity and distribution curves are shown in Fig. 11. At temperatures of 124°C, 166°C, 175°C, 225°C, and 340°C, complete analyses were made, though intensity patterns were obtained at other temperatures, to determine more exactly the temperature at which considerable change in the position of the main peak occurred. This change was found to take place between 157°C and 166°C. In the distribution curves at all temperatures, the nearest

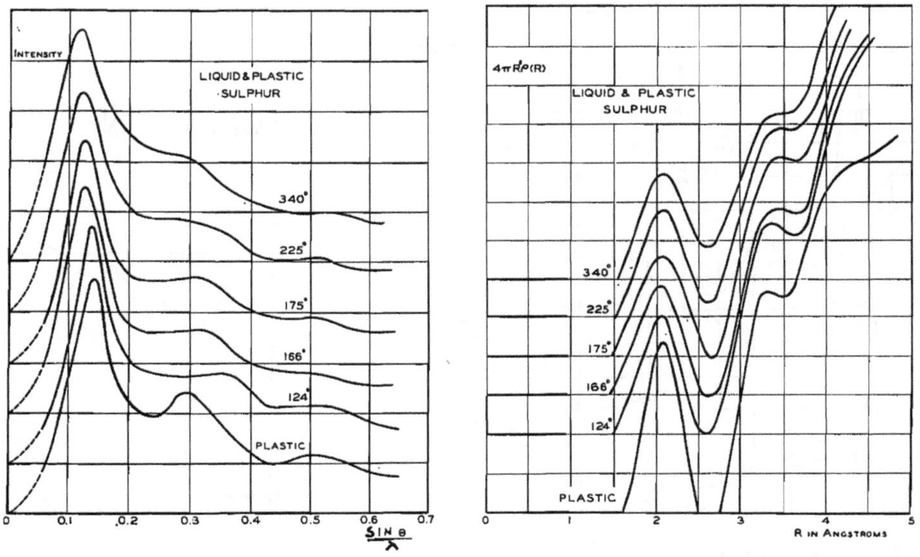

FIG. 11. Intensity curves and distribution curves for liquid sulphur at 124°C, 166°C, 175°C, 225°C, 340°C, and for plastic sulphur at room temperature.

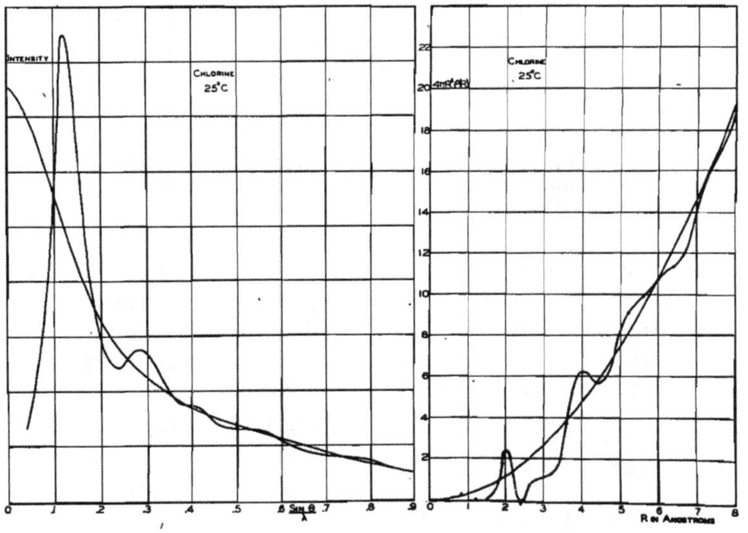

FIG. 12. Intensity curve and distribution curve for liquid chlorine at 25°C and 7.7 atmospheres pressure.

neighbor distance is about 2.07A, and these approximately 1.7 nearest neighbors are indicated as being permanent for temperatures up to about 200°C. It is pointed out that if the S_8 molecule were in the form of an open chain, then the end atoms would have one nearest neighbor, and the intermediate atoms would have two nearest neighbors, giving rise to an average of roughly 1.7 atoms as observed. For plastic sulphur at room temperature, there are two permanent neighbors at 2.08A, characteristic of a closed ring, or of a long chain.

9. Chlorine

Liquid chlorine (b.p. 34.6°C) was investigated by Gamertsfelder[28] at 25°C and under its own vapor pressure of about 7.7 atmospheres. Figure 12 shows the results of this work, with intensity

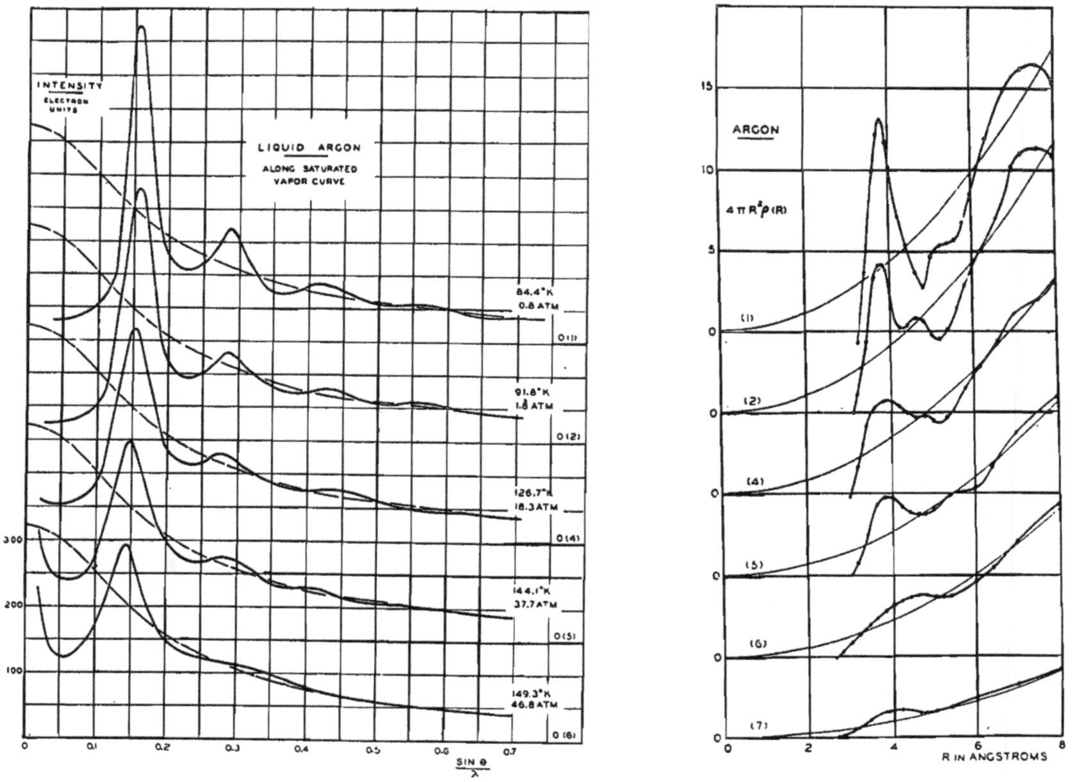

FIG. 13. Intensity curves and distribution curves for liquid argon at 84.4°K and 0.8 atmos.; 91.8°K and 1.8 atmos.; 126.7°K and 18.3 atmos.; 144.1°K and 37.7 atmos.; 149.3°K and 46.8 atmos.; and a distribution curve (No. 7) for argon gas at 149.3°K and 43.8°K.

peaks at $(\sin\theta)/\lambda=0.122$, 0.285, 0.41, 0.55, and 0.8 and with distribution peaks at 2.01A, 2.0A, and 5.2A. The peak at 2.01A represents one (0.97 measured) permanent nearest neighbor, confirming the existence of Cl_2 molecules in the liquid.

10. Argon

In many respects, this element (m.p. 189.2°C; b.p. 185.7°C) is one of the most interesting and suitable to study. Work has been reported by Keesom and de Smedt,[7] Lark-Horovitz and Miller[45] and Eisenstein and Gingrich.[46,47] In the latest work,[47] diffraction patterns have been obtained for the liquid and for the vapor over wide ranges of pressure and temperature. Intensity and distribution curves are shown in Fig. 13 for a few conditions of pressure and temperature, with argon in the liquid state for curves 1, 2, 4, 5, and 6, and in the vapor state for curve 7 (distribution curve only). Many more curves are shown in the original article, particularly for argon vapor, and reference should be made to this article for more detailed results. Referring to the intensity curves of Fig. 13, at least three effects can be observed of the increase in temperature along the saturated vapor curve. The most obvious effect is the progressively greater smearing-out of the pattern; another observation is that the position of the main peak is progressively shifted to smaller angle; and a third effect is the increase in the small angle scattering at high temperatures. This last effect is even more pronounced in the case of high temperature, high pressure vapor, and it is interpreted as being due to the formation of many or of frequently recurring large-scale (10A or greater) density fluctuations, under these conditions. The distribution curves show the most pronounced first peak at low temperatures as expected. Whereas in crystalline argon there are 12 nearest neighbors at 3.82A, in liquid argon at 84.4°K and 0.8 atmos.

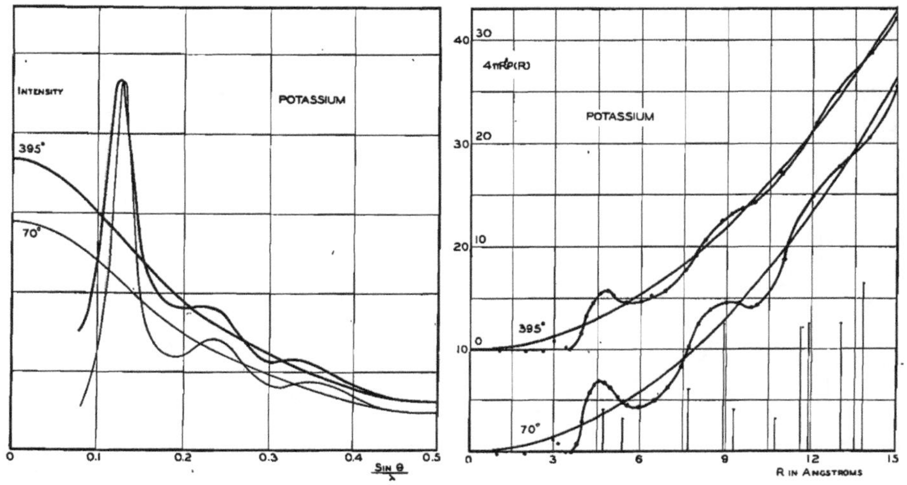

FIG. 14. Intensity curve and distribution curve for liquid potassium.

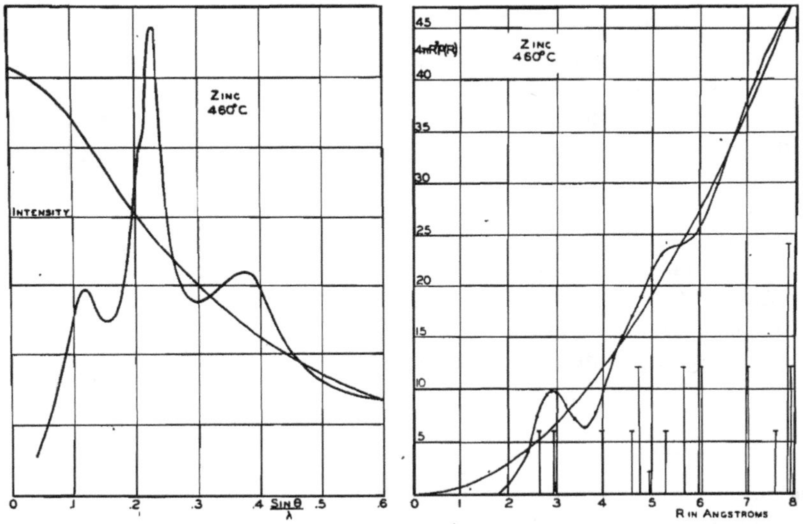

FIG. 15. Intensity curve and distribution curve for liquid zinc at 460°C.

there are 10.2 to 10.9 nearest neighbors at 3.79A. These results show that relatively small changes of temperature have considerable effect on the distribution curve of argon near its triple point.

Theoretical considerations relative to atomic distribution curves have frequently made reference to liquid argon. Details of the several approaches to this problem and of the results so far obtained may be found in the work, for example, of Wall,[36] Rice,[48] Rushbrooke and Coulson,[39] Rushbrooke,[49] Corner and Lennard-Jones,[50] Kirkwood,[51] and Kirkwood and Boggs.[52]

11. Potassium

Liquid potassium (m.p. 62.3°C) has been studied by Keesom[32] and by Randall and Rooksby[33] at one temperature just above the melting point, and by Thomas and Gingrich[53] at the two temperatures 70°C and 395°C. Curves from this latter work are shown in Fig. 14. At 70°C intensity peaks occur at $(\sin\theta)/\lambda = 0.130$, 0.233, 0.36, whereas at 395°C peaks were found at 0.126, 0.22, and 0.35. In the distribution curves, the first peak is at 4.64A for 70°C and at 4.76A for 395°C, with both peaks representing an average concentration

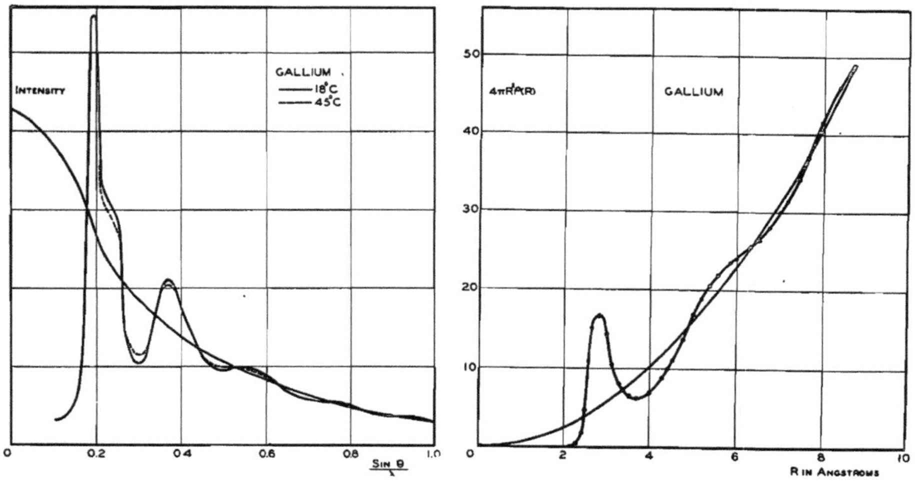

FIG. 16. Intensity curves for liquid gallium at 18°C and 45°C and a distribution curve for liquid gallium at 18°C.

of about 8 atoms. These results have been used by de Boer and Michels,[54] Hildebrand,[55] and Gingrich and Wall[56] in their theoretical discussions relative to atomic distributions. In the latter case, for example, the latent heats of fusion and of vaporization of potassium have been calculated with some success, using the distribution curves shown here. Hildebrand has also calculated from these distribution curves the ratio of the energy of vaporization at these two temperatures, obtaining surprisingly close agreement with that from the vapor pressure curve.

12. Zinc

The only investigation on liquid zinc (m.p. 419°C) has been made by Gamertsfelder[28] at 460°C. Figure 15 shows the intensity and distribution curves in this case. As in the case of aluminum, the first intensity peak is weak, and the second peak is the most intense. The intensity peaks occur at $(\sin\theta)/\lambda = 0.115$, 0.228, 0.38. The first distribution peak at 2.94A represents an average concentration of about 10.8 atoms.

13. Gallium

Liquid gallium (m.p. about 30°C, and easily undercooled) has been studied by Menke[57] at 18° and at 45°C. Results of this work are shown in Fig. 16. Intensity peaks occur at $(\sin\theta)/\lambda = 0.195$, 0.372, 0.57, 0.77, and 0.96, and distribution peaks occur at 2.83A and 5.8A. It is to be noted that there is but slight difference between the intensity patterns at the two temperatures in spite of the fact that at 18°C, the liquid is undercooled.

14. Selenium

The only reported work on liquid selenium (m.p. 217°C) is that of Prins[43] who simply lists "spacings" equivalent to intensity peaks at $(\sin\theta)/\lambda = 0.148$, 0.280, 0.435, states that there was no appreciable effect of temperature on the pattern, and notes that the pattern was indentical with one obtained from a preparation quenched by dropping into water. Lark-Horovitz and Miller[58] have worked with amorphous selenium, obtaining intensity peaks at $(\sin\theta)/\lambda = 0.146$, 0.289, 0.443 and distribution peaks at 2.35A, 3.7A, and 4.8A.

15. Rubidium

Liquid rubidium (m.p. 38.4°C) has been reported by Randall and Rooksby[33] as giving a diffraction pattern with an intensity peak at $(\sin\theta)/\lambda = 0.122$

16. Cadmium

Gamertsfelder[28] has worked with liquid cadmium (m.p. 321°C) at 350°C. His results are shown in Fig. 17. Intensity peaks occur at

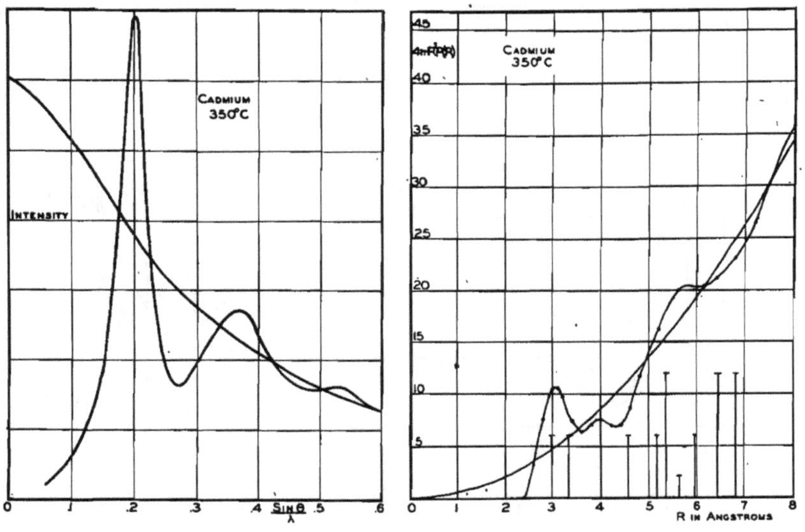

FIG. 17. Intensity curve and distribution curve for liquid cadmium at 350°C.

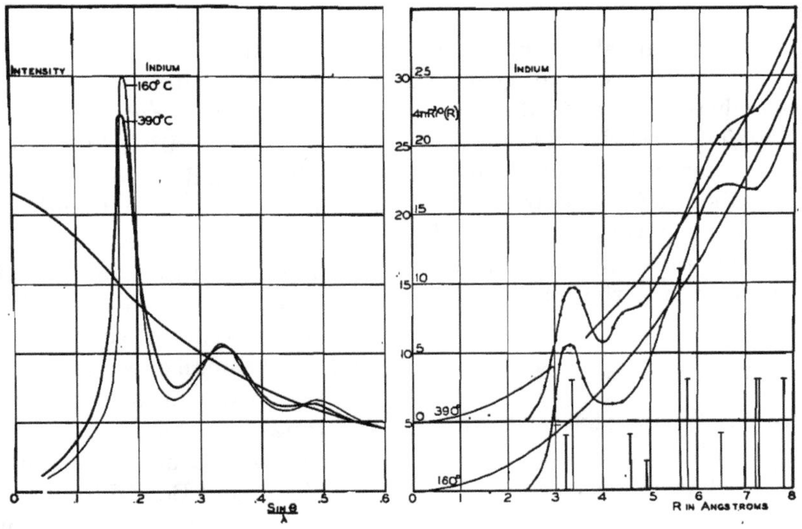

FIG. 18. Intensity curves and distribution curves for liquid indium at 160°C and 390°C.

$(\sin\theta)/\lambda = 0.203$, 0.360, 0.55 and the first distribution peak, representing an average concentration of 8.3 atoms, is at 3.06A.

17. Indium

Liquid indium (m.p. 155°C) has been studied by Gamertsfelder[28] at 160°C and at 390°C. The results of his work are shown in Fig. 18. The intensity peaks for 160°C are at $(\sin\theta)/\lambda = 0.179$, 0.335, 0.50 and at 390°C they occur at 0.177, 0.335, 0.49. The first distribution peak at 160°C represents about 8.5 atoms at an average distance 3.30A, while that at 390°C represents about 8.4 atoms at roughly 3.36A.

18. Tin

Liquid tin (m.p. 231.9°C) has been dealt with by Sauerwald and Teske,[59] Prins,[43] Danilov and Radtschenko,[60] and Gamertsfelder.[28] In the latter case, temperatures of 250°C and 390°C

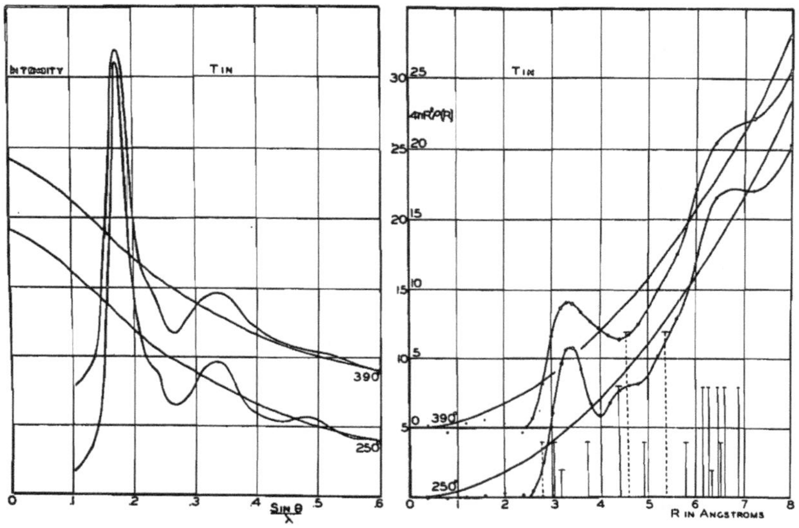

FIG. 19. Intensity curves and distribution curves for liquid tin at 250°C and 390°C.

were used and these results are shown in Fig. 19. At 250°C, intensity peaks occur at $(\sin\theta)/\lambda = 0.174$, 0.335, 0.49 whereas at 390°C they occur at 0.175, 0.330, 0.50. The first distribution peak for tin at 250°C represents about 10 atoms at 3.38A, and at 390°C, about 8.9 atoms at 3.36A. It is interesting to note from this work that the liquid structure is much more analogous to that of crystalline white tin than to that of crystalline gray tin.

19. Cesium

Randall and Rooksby[33] report having found an intensity peak in the diffraction pattern of liquid cesium (m.p. 28.5°C) at $(\sin\theta)/\lambda = 0.111$.

20. Mercury

Liquid mercury (m.p. 38.9°C) has been investigated by Prins,[61] Coster and Prins,[62] Wolf,[63] Raman and Sogani,[64] Debye and Menke,[9,65] Sauerwald and Teske,[59] Menke,[57] and Boyd and Wakeham.[66] Most of these investigations led to interesting qualitative results, but the investigations of Debye and Menke,[9,65] Menke,[57] and Boyd and Wakeham[66] have supplied more detailed information. The results of Debye and Menke[9] are shown in Fig. 20 for mercury at room temperature. Intensity peaks occur at $(\sin\theta)/\lambda = 0.162$, 0.328, 0.52, 0.68, 0.82, and distribution peaks occur at 3.23A and 6.5A. Using the reflection method, as did Debye and Menke, Boyd and Wakeham[66] reported work for temperatures of −36°C, −34°C, 0°C, 30°C, 75°C, 125°C, 175°C, and 250°C, and, among other differences from previous work, they found an extra intensity peak at small angle. Unfortunately, however, lack of purity of their radiation raises some doubt as to the origin of this peak.[15] Nevertheless, the possible existence of this extra peak should have a relatively small effect on the distribution curve, and their work does supply information for the first time concerning the temperature effect upon the distribution curve for mercury. Hildebrand, Wakeham, and Boyd[67] have used these results to calculate the intermolecular potential in the case of mercury.

21. Thallium

Sauerwald and Teske[59] report the existence of intensity peaks in the diffraction pattern of liquid thallium (m.p. 303.5°C) at $(\sin\theta)/\lambda = 0.164$, 0.323, 0.442.

22. Lead

Randall and Rooksby[68] report an intensity peak for liquid lead (m.p. 327.5°C) at $(\sin\theta)/\lambda = 0.173$.

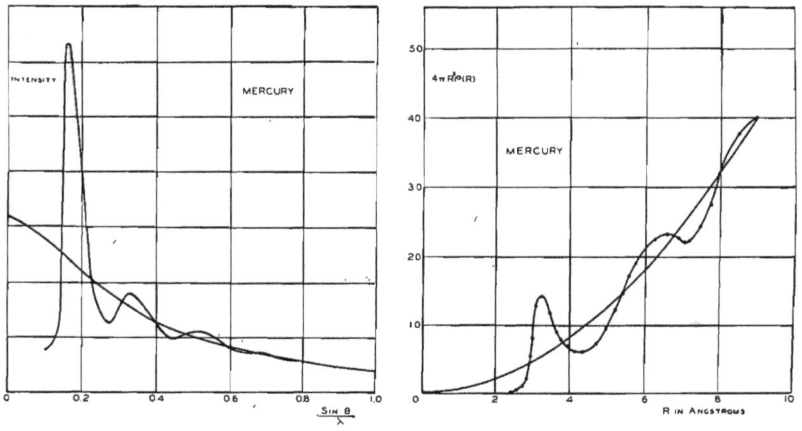

FIG. 20. Intensity curve and distribution curve for liquid mercury at room temperature.

23. Bismuth

Randall and Rooksby,[68] Sauerwald and Teske[59] and Danilov and Radtschenko[60] report the position of the main intensity peak for liquid bismuth (m.p. 271°C). The average of these three somewhat divergent values is $(\sin\theta)/\lambda = 0.164$.

V. SUMMARY

Experimental work on the diffraction of x-rays by liquid elements has been reported for 23 elements. These determinations in and by themselves are valuable for their contribution to the totality of our knowledge of this particular physical phenomenon. The list of elements for which this information is known should be extended to cover all elements, and it is to be hoped that more complete determinations of the individual diffraction patterns can be given as advances are made in experimental techniques. But a more valuable contribution of this type of work is in making available directly determined descriptions of the structure of various liquid elements, which structure can be correlated with, or used to predict physical properties of the elements.

The structures of liquid elements in terms of the atomic distribution curves have been reported for 16 elements, ten of which reports include determinations of the effect of temperature, and one of which includes the effect of pressure as well. These atomic distribution curves represent the time-averaged atomic environment about any given atom within the liquid, but this environment is neither permanent, as in a crystal, nor random, as in a gas, and hence no simpler description of liquid structures can now be given to supply the same information. These first 16 determinations of atomic distribution curves constitute the beginning of a complete list of all elements, to parallel a list of the crystal structures of the elements, as an aid to more comprehensive understanding of the structure of matter and of the transitions of matter from one state to another.

Reference has been made, in the present review, to some of the initial attempts toward securing more comprehensive understanding of the structure of matter with the aid of work here reported. On the basis of assumed models of the liquid state, distribution curves have been computed and compared with the observed curves. From these observed distribution curves, interatomic potentials have been deduced, and physical properties of the elements, such as latent heats of fusion and vaporization have been calculated, with some success. In addition to this, the distribution curves have supplied very direct evidence to confirm the existence of molecules in some liquid elements (e.g., N_2, O_2, Cl_2, P_4) and to imply the possible existence of more complicated atomic aggregates in a few cases.

REFERENCES

(1) W. Friedrich, P. Knipping, and M. v. Laue, Sitzb. Math.-Phys. Klasse Bayer. Akad. Wiss. Munchen 303 (1912).

(2) See P. P. Ewald, *Kristalle und Rontgenstrahlen.* (J. Springer, Berlin, 1923); *Handbuch der Physik* (1933), second edition, Vol. 23; R. W. G. Wyckoff, *The Structure of Crystals* (Chemical Catalog Company, New York, 1931).

(3) C. G. Barkla and T. Ayers, Phil. Mag. **21**, 275 (1911).

(4) E. O. Wollan, Rev. Mod. Phys. **4**, 205 (1932).

(5) W. Friedrich, Physik. Zeits. **14**, 397 (1913).

(6) P. Debye and P. Scherrer, Gottingen Nachrichten 16 (1916).

(7) W. H. Keesom and J. de Smedt, Proc. Amst. Akad. Sci. **25**, 118 (1922); **26**, 112 (1923).

(8) F. Zernicke and J. Prins, Zeits. f. Physik **41**, 184 (1927).

(9) P. Debye and H. Menke, Ergeb. d. Tech. Rontgenk. II (1931).

(10) See J. T. Randall, *The Diffraction of X-Rays and Electrons by Amorphous Solids, Liquids and Gases* (John Wiley and Sons, Inc., New York, 1934), p. 107.

(11) P. Debye, Ann. d. Physik **46**, 809 (1915).

(12) B. E. Warren and N. S. Gingrich, Phys. Rev. **46**, 368 (1934).

(13) B. E. Warren, J. App. Phys. **8**, 645 (1937).

(14) A. H. Compton and S. K. Allison, *X-Rays in Theory and Experiment* (D. Van Nostrand Company, New York, 1935), p. 116.

(15) R. Q. Gregg and N. S. Gingrich, Rev. Sci. Inst. **11**, 305 (1940).

(16) C. Gamertsfelder and N. S. Gingrich, Rev. Sci. Inst. **9**, 154 (1938).

(17) A. Eisenstein and N. S. Gingrich, Rev. Sci. Inst. **12**, 582 (1941).

(18) F. C. Blake, Rev. Mod. Phys. **5**, 180 (1933).

(19) G. P. Mitchell, unpublished work done at the University of Missouri.

(20) See reference 14, p. 781.

(21) See reference 14, p. 782.

(22) N. S. Gingrich, Phys. Rev. **59**, 290 (1941).

(23) R. Hultgren, N. S. Gingrich, and B. E. Warren, J. Chem. Phys. **3**, 351 (1935).

(24) See J. W. Mellor, *Higher Mathematics for Students of Chemistry and Physics* (Longmans, Green and Company, 1929), p. 469.

(25) G. C. Danielson and C. Lanczos, J. Frank. Inst. **233**, 365 (1942); **233**, 435 (1942).

(26) W. H. Keesom and K. W. Taconis, Physica **4**, 28 (1937); **4**, 256 (1937); **5**, 270 (1938); Proc. Amst. Akad. Sci. **41**, 194 (1938).

(27) J. Prins and H. Petersen, Physica **3**, 147 (1936).

(28) C. Gamertsfelder, J. Chem. Phys. **9**, 450 (1941).

(29) P. C. Sharrah, Ph.D. Dissertation, University of Missouri, Columbia, Missouri (August, 1942).

(30) G. G. Harvey, Phys. Rev. **46**, 441 (1934).

(31) P. C. Sharrah and N. S. Gingrich, J. Chem. Phys. **10**, 504 (1942).

(32) W. H. Keesom, Proc. Amst. Akad. Sci. **30**, 341 (1927).

(33) J. T. Randall and H. P. Rooksby, Nature **130**, 473 (1932).

(34) L. P. Tarasov and B. E. Warren, J. Chem. Phys. **4**, 236 (1936).

(35) F. H. Trimble and N. S. Gingrich, Phys. Rev. **53**, 278 (1938).

(36) C. N. Wall, Phys. Rev. **54**, 1062 (1938).

(37) Reference 10, p. 134.

(38) C. D. Thomas and N. S. Gingrich, J. Chem. Phys. **6**, 659 (1938).

(39) G. S. Rushbrooke and C. A. Coulson, Phys. Rev. **56**, 1216 (1939).

(40) A. H. Blatchford, Proc. Phys. Soc. London **45**, 493 (1933).

(41) S. R. Das, Ind. J. Phys. **12**, 163 (1938).

(42) S. R. Das and K. Das Gupta, Nature **143**, 332 (1939).

(43) J. A. Prins, Trans. Faraday Soc. **33**, 110 (1937).

(44) N. S. Gingrich, J. Chem. Phys. **8**, 29 (1940).

(45) K. Lark-Horovitz and E. P. Miller, Nature **146**, 459 (1940).

(46) A. Eisenstein and N. S. Gingrich, Phys. Rev. **58**, 307 (1940).

(47) A. Eisenstein and N. S. Gingrich, Phys. Rev. **62**, 261 (1942).

(48) O. K. Rice, J. Chem. Phys. **7**, 883 (1939).

(49) G. S. Rushbrooke, Proc. Roy. Soc. Edinburgh **60**, 182 (1940).

(50) J. Corner and J. E. Lennard-Jones, Proc. Roy. Soc. London **A178**, 401 (1941).

(51) J. G. Kirkwood, J. Chem. Phys. **7**, 919 (1939).

(52) J. G. Kirkwood and E. M. Boggs, J. Chem. Phys. **10**, 394 (1942).

(53) C. D. Thomas and N. S. Gingrich, J. Chem. Phys. **6**, 411 (1938).

(54) J. de Boer and A. Michels, Physica **6**, 97 (1939).

(55) J. H. Hildebrand, J. Chem. Phys. **7**, 1 (1939).

(56) N. S. Gingrich and C. N. Wall, Phys. Rev. **56**, 336 (1939).

(57) H. Menke, Physik. Zeits. **33**, 593 (1932).

(58) K. Lark-Horovitz and E. P. Miller, Phys. Rev. **51**, 380 (1937).

(59) F. Sauerwald and W. Teske, Zeits. f. anorg. allgem. Chemie **210**, 247 (1933).

(60) J. V. Danilov and V. J. Radtschenko, Physik. Zeits. Sowjetunion **12**, 745 (1938).

(61) J. A. Prins, Physica **6**, 315 (1926).

(62) D. Coster and J. A. Prins, J. de phys. et rad. **9**, 153 (1928).

(63) M. Wolf, Zeits. f. Physik **53**, 72 (1929).

(64) C. V. Raman and C. M. Sogani, Nature **120**, 514 (1927).

(65) P. Debye and H. Menke, Physik. Zeits. **31**, 797 (1930).

(66) R. N. Boyd and H. R. Wakeham, J. Chem. Phys. **7**, 958 (1939).

(67) J. H. Hildebrand, H. R. Wakeham, and R. N. Boyd, J. Chem. Phys. **7**, 1094 (1939).

(68) J. T. Randall and H. P. Rooksby, Trans. Faraday Soc. **33**, 109 (1937).

Printed in the USA
CPSIA information can be obtained
at www.ICGtesting.com
LVHW011927071123
763206LV00005B/58

9 781258 662134